Designing Human-centred Technology
A Cross-disciplinary Project in Computer-aided Manufacturing

The Springer Series on

ARTIFICIAL INTELLIGENCE AND SOCIETY

Series Editor: KARAMJIT S. GILL

Knowledge, Skill and Artificial Intelligence
Bo Göranzon and Ingela Josefson (Eds.)

Artificial Intelligence, Culture and Language: On Education and
Work
Bo Göranzon and Magnus Florin (Eds.)

Designing Human-centred Technology: A Cross-disciplinary
Project in Computer-aided Manufacturing
H. H. Rosenbrock (Ed.)

H. H. Rosenbrock (Ed.)

Designing Human-centred Technology

A Cross-disciplinary Project in Computer-aided Manufacturing

With 18 Figures

Springer-Verlag London Ltd.

Howard Rosenbrock
Linden, Walford Road, Ross-on-Wye,
Herefordshire HR9 5PQ

Cover illustration adapted from Fig. 2.2. The evolution of the lathe in the early nineteenth century.

ISBN 978-3-540-19567-2   ISBN 978-1-4471-1717-9 (eBook)
DOI 10.1007/978-1-4471-1717-9

British Library Cataloguing in Publication Data
Rosenbrock, H. H. (Howard H, 1920–)
    Designing human-centred technology.
    1. Manufacture. Applications of computer systems
    I. Title
    670.42′7

Library of Congress Cataloging-in-Publication Data
Designing human-centred technology: a cross-disciplinary project in
    computer-aided manufacturing/H. H. Rosenbrock, ed.
        p. cm.—(The Springer series on Artificial Intelligence and Society)
    Includes bibliographical references.

    1. Computer integrated manufacturing systems.   2. Human engineering.
    I. Rosenbrock, H. H.   II. Series
    TS 155.6.D474   1989
    670′.285—dc20   89-21625
                                                         CIP

Filmset by MJS Publications Ltd, Buntingford, Herts.

2128/3916–543210 (Printed on acid-free paper)

Foreword

This second book in our series *Artificial Intelligence and Society* explores the issues involved in the design and application of human-centred systems in the manufacturing area. At first glance it may appear that a book on this topic is somewhat peripheral to the main concerns of the series. In fact, although starting from an engineering perspective, the book addresses some of the pivotal issues confronting those who apply new technology in general and artificial intelligence (AI) systems in particular. Above all, the book invites us to consider whether the present applications of technology are such as to make the best use of human skill and ingenuity and at the same time provide for realistic and economically sustainable systems design solutions. To do so it is necessary to provide systems which support the skill, and are amenable to the cultures, of the areas of application in question. In a philosophical sense it means providing tools to support skills rather than machines which replace them, to use Heidegger's distinction.

The book gives an authoritative account of the University of Manchester Institute of Science and Technology (UMIST) tradition of human-centredness and provides a participatory design approach which focuses on collaborative learning and enhancement and creation of new skills. It also argues that collaboration should be supported by institutions through the creation of supportive infrastructures and research environments. It emphasises the optimisation of practical knowledge with the help of scientific knowledge and rejects the alternative. The 'blank table debate' in the book shows how personal opinions, professional experiences, academic and practical perspectives of a multi-disciplinary team of designers can work together creatively, resolve conflicts through a constructive divergence and arrive at practical systems which are of benefit to the users. The book records and analyses the reasons for the research conflicts which are believed to be a product of the institutional splitting which an interdisciplinary project faces.

The book raises, both directly and indirectly, issues of central concern to the present series. Among the series' objectives are:

it will provide an authoritative treatment of the nature of knowledge, skill and practice and the nature of AI technology, and will examine issues of the design, application and implications of AI systems within their broader context. The series will focus on the philosophical, social, economic and political contexts of AI development and will cover topics such as the nature of expert knowledge and the problems of its computerisation; AI technology and the transfer of knowledge and skills; human/computer integration in the management of industrial and military systems; social and cultural shaping of AI technology and the nature of work; knowledge-based society and the issues of control, responsibility, access and ethics. (*AI & Society* pamphlet.)

Given the growing significance of computerisation in all aspects of human endeavour, and the anticipated pervasive nature of AI systems, it behoves each professional group to explore creatively the issues and implications of technological change in their own areas. In the field of manufacturing technology it is becoming clear that many sections of industry are already precariously dependent upon machine-centred systems. Such systems are typically highly synchronised and coordinated, but frequently lack robustness, in the sense that they are inadequate in dealing with disturbances and uncertainties. Thus, if one part of the system goes down, the high level of synchronisation is suddenly turned into its opposite and becomes a high level of desynchronisation. If, however, the system is human-centred as suggested in this book, it is contended that this will provide for good human–machine symbiosis, in which there will be involved pro-active, creative human beings for whom the system is transparent, and who will be capable of acting in an informed way in the event of uncertainty.

We are at an interesting juncture in the development of AI which could prove to be a turning point for the design of intelligent systems. Dominant AI is rapidly finding its place in software engineering or cognitive science which, by nature and by adoption, have a machine-centred focus exemplified by the expert systems and laboratory-based human–computer interaction research. This book provides an alternative to the computational metaphor of AI and recognises the significance of the tacit dimension of human knowledge, and demonstrates the significance of working-life experiences and practices for the design of purposeful computer systems. In this context it is consonant with the increasing interest of industry, commerce and organisations which are beginning to recognise the limitations of machine-centred technology to assist in qualitative decision-making processes.

The introduction of AI systems into any field of skill or competence must reflect in part the nature of existing practices and technology. If the practices which precede the introduction of the AI systems are such as to diminish the role of the human beings and in many cases reduce them to abject machine appendages, it will tend to follow that AI systems will reflect these values, and an opportunity for a more creative and ultimately productive system will have been lost. The systems will tend to be "expert replacement systems" rather than "expert systems". Present systems are introduced within a scientific tradition in which it is held that a system is only scientifically designed if it displays the three predominant characteristics of Western science and technology: predictability, repeatability and mathematical quantifiability. That by definition precludes intuition, subjective judgement and tacit knowledge. A human-centred approach questions, at a philosophical level, the wisdom of such an approach, and suggests rather one in which the best use is made of the human being and the system, paying due regard to the characteristics of both.

This book clarifies the debate on the tacit dimension of skills and practice through the UMIST project and thus lays a foundation for the design of practical systems in working-life environments where the fundamental issues of collaborative design will be common. It will be a challenge to AI researchers who may wish to transfer these methodologies to their own domains such as collaborative learning, knowledge transfer, office automation, human–computer interaction, decision support, natural language processing. The book will be of special interest to researchers and practitioners who are involved in the design of collaborative knowledge-based networks, multi-media learning systems, and collaborative training systems. The design challenge may be seen in dealing with the issues such as: sharing of knowledge space; sharing common experiences, distribution and diffusion of knowledge; open and distance learning; tacit knowledge; social and cultural determinants; human factors; information flows and knowledge flows; internationalisation of knowledge – transfer of knowledge across cultures; and mediation of knowledge. This book provides concepts, approaches and methodologies which may be transferred to the design of AI systems in the above areas and in dealing with the above issues. It shows how the UMIST tradition of human–machine symbiosis has evolved a methodology for human-centred design during the process of the project. This methodology became an important part of the ESPRIT project 1217. It will be reasonable to suggest that the concept of human–machine symbiosis can be applied equally to develop participatory design methodologies for domains other than manufacturing.

This book concludes with a brief outline of ESPRIT project 1217 – Human-centred CIM, which was greatly influenced by the UMIST research. Other significant influences included the social shaping philosophy of the German partners and the end user involvement techniques of the Danish partners. A subsequent book in this series will explore these influences and will describe the ESPRIT project in greater detail.

A great strength of the book is that it raises at a practical and a philosophical level sets of alternatives to the given orthodoxy. As such it challenges and invites other professions, such as those of law, medicine and banking, likewise to explore the underlying values of the systems being introduced in their areas, and to question whether the new technologies, including expert systems and AI, might not be introduced in such a way as to enhance human skill rather than diminish it.

Mike Cooley
Karamjit S. Gill

Preface

The Industrial Revolution of the late eighteenth and early nineteenth centuries took place under social conditions very different from our own. One of its legacies is a technology relying upon assumptions which are strongly at variance with the needs and aspirations of men and women today. Production systems are designed with attention fixed upon the machines: their needs and their effective use are the main considerations. Men and women have to fit as best they can into the systems that result. Often they are subordinated to the machines, which determine human actions and rate of working. A specifically human input – by initiative, skill, control, and the response to unexpected events – is rejected.

This situation has been strongly criticised by social scientists for its damaging effects upon workers. It can equally be criticised for its wastefulness of a primary resource. It makes excellent use of machines, but very poor use of people. Systems designed to reject a human input, or to draw this only from a small centralised nucleus of people, are rigid and inflexible. They cannot respond to unforeseen events, or to rapidly changing market demands. So much is now becoming acknowledged.

If we wish to change the direction of technological development, so that the human input is accepted and valued, we shall have to intervene at the stage where new technology is being designed. The project at the University of Manchester Institute of Science and Technology (UMIST) which is described in this book was an attempt to do so in relation to new production systems which go under various names, but are usually known collectively as computer-integrated manufacturing (CIM). The technology, still fluid and developing, sets out to control and integrate the whole production process by means of a network of computers. As usually envisaged, the aim is to replace every human contribution by automation, with the ultimate goal of a 'workerless factory'. The dream is an old one, and it is well known that the result is not a factory without workers, but a factory in which workers must behave like machines.

In the project a different aim was proposed: computers and automation would not be used to reject human abilities and skill, but rather to cooperate with them to make them more productive. People should not be subordinate to machines; machines should be subordinate to people. Existing skills should be accepted, and room for their exercise should be provided. But room also should be provided for skills to change as technology changes: the evolution of skill was regarded as open-ended and unlimited, where the conventional view sees past skills as finite and doomed to extinction as they are replaced by automation. It was believed that, by following the proposed aim, a system could be designed that was at least as economic as conventional systems, but more flexible, more responsive, and better matched to the aspirations of workers.

One of the major obstacles to the proposed development is the burden of preconceptions brought to the design process by technologists – engineers, computer scientists and others. Their training and experience tend very strongly to concentrate their attention on machines, rather than people. On the other hand, there is a strong tradition in the social sciences of 'socio-technical design', in which the technology and the conditions of work are both studied together in the design stage. Unfortunately, opportunities to do this have usually been limited to small changes at the fringe of technology, rather than the deep reconsideration which is needed. A way of describing the aims of the project is that it sets out, by means of a collaboration between social scientists and technologists, to apply socio-technical design principles at a deep level in technology.

Such an aim poses great problems to those concerned, both technologists and social scientists. The technologists have no experience of designing with the human requirements in view. The social scientists have no deep understanding of the technology, or of the technological design process. So engineers and computer scientists cannot see how to balance human requirements against those of the technology. Social scientists cannot see when a stated technical requirement represents an engineering constraint, and when it is simply a preconception. On both sides it is necessary to acquire the knowledge and practical ability needed to carry out a new kind of design leading to a new kind of technology.

Equilibrium could probably have been achieved in two ways: with social scientists acting as advisers to technologists, but at arm's length from the design process (a not uncommon arrangement), or conversely with social scientists controlling the project and making the major decisions. Neither extreme was thought likely to be fruitful. Instead a tension between the two sides was sought, out of which it was hoped that a joint understanding could arise. The account which follows shows what happened in this process, as seen from different points of view. Disagreements of opinion are as likely to be informative as agreements, and they have not been disguised: the account has the nature of 'Faithful Contendings Displayed'.

With this aim in mind, no attempt has been made to achieve uniformity in the style or approach of the different contributions. To change an author's style is to make a subtle change in the impression conveyed, and to that extent to falsify the account. So, for example, the different practice of authors in regard to the masculine bias in the English language has been left unchanged. The majority have adhered to the traditional convention that 'he' includes 'she', whereas in other places 'he or she' or some other form has been used.

Some may conclude from a reading of what follows that my own role, as originator and grant-holder for the project, called for a more decisive direction of the work. Certainly this would have resulted in greater progress in the technological design, but it would have left many things less clearly defined – as agreements or disagreements or as open questions – than they will be found to be here. My aim was to elicit the views of others; and to state my own clearly and forcibly, but to refrain either from imposing them or from abandoning them if I was unconvinced. The reader should therefore find much material here for study and further development.

What is most to be desired is that the small beginnings in this project should be followed by a multitude of others of increasing depth and richness. The ESPRIT project, described in Chapter 11, carries on and broadens the direct line of the UMIST project. Similar work is being done elsewhere, and extensions in many directions are possible: from batch production to mass production, to office work, to professional work in relation to 'expert systems', and so on. The intellectual basis of 'human-centred' systems needs further development, while as the early attempts to achieve these systems come into use their success in human and economic terms will need to be studied, analysed and learned from.

If those who engage in this work can achieve a change in the direction in which our technology develops, towards something which enriches the human spirit rather than impoverishing it, they will have the gratitude and thanks of those whose work is described here, but also and more importantly of a great multitude who will work with that more 'human-centred' technology.

Ross-on-Wye   *Howard Rosenbrock*
1 November 1988

Contents

Abbreviations XVI

1. The Background to the Project
Howard Rosenbrock 1

Introduction 1
Why is Taylorism so Widespread? 3
An Alternative Technology? 4
The Steering Committee 6
The Application for Funding 8
Steering Committee Discussions 9
Methodology 10
The Aims of the Project 12

2. The Technical Problem
Howard Rosenbrock 17

Introduction 17
Some History 17
Numerical Control 22
Record–Playback 24
Cutting Technology 27
Design 28
Learning, but Learning What? 30
My Own View 32
Terminology 35

3. Towards a Design Methodology: A Psychologist's View
Martin Corbett 37

Introduction 37
The Design Context 37
Design Criteria Related to the Human as a Component of the
  System 39
Design Criteria Related to the System as the Operator's
  Environment 39
Heuristics in Design Negotiation: The 'Blank Table Debate'... 41
Towards an Interactive Design Process 42
Conclusion 43
Appendix: Operator Control and Automation
Howard Rosenbrock and Martin Corbett 44

4. The Social and Engineering Design of Computer Numerically Controlled Technology
Paul Kidd 51

Introduction 51
Numerical Control Technology 52
The Development of Cutting Technology Software 55
The 'Blank Table Debate' – A Divergence of Opinion 61
Conclusions and Personal Reflections – The Engineer as a
   Social Scientist 63

5. A Computer Science View
Roger Holden 65

Introduction 65
Software Structure 66
Aspects of the UMIST System 69
Software Development 72
Use of Alternative Programming Methods 75
Conclusion 78
Appendix: Skeleton Syntax for Part-programs 79

6. On the Collaboration Between Social Scientists and Engineers
Lisl Klein 81

Background – The Dynamics 81
Models of Science 85
Operational Issues 91
Making It Happen – Institutions and Infrastructure 95

7. (How) Can Technology be Redirected? A Scandinavian Perspective
Håkon Finne 99

Introduction 99
The Concept of Non-subordinating Machinery 100
Methodologies for Designing Redirected Technology 105
Was the Project Appropriate and Efficacious in a Redirection
   Effort? 108
Future Developments 112

8. A Works Director's View
Allan Chatterton 117

Introduction 117
Evolution of New Technology and Human Skills 118
The Operator, His Machine, and a Human-centred Approach
   to Technology 119
The UMIST Project 120
Conclusion 121

9. The Coordinator's View
Harold Palmer 123

Introduction 123
General Achievements 123
The Conduct of the Project 124
Cross-disciplinary Work by Engineers and Social Scientists.. 125
The Importance of Management and of Technical Process
    Development in Human-centred Working 126
The Future 131
Final Comment 131

10. Human-centred Systems
Mike Cooley 133

Introduction 133
The Early Stages 133
Technology, Science and Ideology 134
Proposals for a New Approach 135
First Practical Moves 136
Issues of Status, Skill and Uncertainty 137
Human-centred Capabilities 140
Future Development of Human-centredness 141

11. The ESPRIT Project
Shaun Murphy 145

Introduction and Summary 145
The Meaning of Human-centredness 146
The Factory of the Future 147
ESPRIT Project 1217 (1199) 152
CIM Design Considerations 153
The BITZ Demonstration Site 153
The Shop-floor Monitor and Controller 155
The Sketching Module 156
The BICC Demonstration Site 157
The Rolls Royce Demonstration Site 164
The Human-centred Lathe Controller 165
Concluding Remarks 167
Appendix: Participating Organisations 168
Glossary 168

12. Postscript
Howard Rosenbrock 169

Appendices 173
1. Short CVs of Contributors 173

2. SERC Application 177

3. Transcript of a Part of the Steering Committee Meeting
    Held on 12 July 1982 185

Abbreviations

AI	artificial intelligence
BITZ	Bremer Innovations und Technologiezentrum
CAD	computer-aided design
CAITS	Centre for Alternative Industrial and Technological Systems
CAM	computer-aided manufacture
CAP	computer-aided planning
CIM	computer-integrated manufacture
CNC	computer numerically controlled
CV	curriculum vitae
DNC	direct numerical control
EEC	European Economic Community
FMS	flexible manufacturing system
MDI	manual data input
MIT	Massachusetts Institute of Technology
NC	numerical control
PR	public relations
R & D	research and development
RA	research assistant
RPM	revolutions per minute
SERC	Science and Engineering Research Council
SSRC	Social Science Research Council
SMC	shop-floor monitor and controller
UMIST	University of Manchester Institute of Science and Technology
USAF	United States Air Force

Further abbreviations will be found in the Glossary for Chapter 11, on page 168

Chapter 1

The Background to the Project

Howard Rosenbrock

Introduction

The personal experience out of which this project grew can be described in
very few words. From about 1968, in collaboration with colleagues, I had
been developing a theory, and a large software package, which would allow
the computer-aided design (CAD) of control systems. The aim was to give
the designer a more powerful tool to aid him in his work: a computer with
graphic display and interactive conversational working, rather than paper
and pencil and slide-rule. Our approach at that time was not a question of
principle, but arose from previous industrial experience and our intention to
use the methods we developed for our own purposes.

By 1974 it was coming to be accepted that this approach was successful. It
allowed the designer to investigate a tentative design, and to see whether it
was stable, what was the margin of stability, the speed of response, and so
on. If these were unsatisfactory, it suggested ways in which they could be
improved. An existing skill in the designer was presupposed, and for the
most effective use of the CAD system this had to develop in certain rather
natural directions.

If this work was successful at one level, it was quite unsuccessful at
another. For many years, the preferred lines of research in control engineering had been very different. They had been aimed not at accepting the
designer's skill and trying to extend it and make it more productive but
rather at replacing his skill by a large computer-algorithm. The designer, it
was supposed, would specify the problem in great detail, including, say, the
dynamic behaviour of the plant to be controlled, the constraints, and a cost
function to be minimised. Then the computer would find the best solution
satisfying these requirements.

There were some technical reasons for this tendency. In the previous decade, interactive computing was not widely available, and methods were adapted to off-line batch working. The aerospace problem, then heavily funded, was also better suited than most industrial control problems to an algorithmic approach. There was clearly more than this involved, however, because objections were raised to the need for skill in the user. What we had regarded as an advantage of our approach, its use of an existing skill and the opportunity it gave for this skill to develop further, was seen as a defect: a method which would work only by virtue of the skill of its user was seen as, to that extent, incomplete.

Engineering problems can seldom be reduced to two independent stages, of formulation followed by solution. Consequently, the actual practice of designers using algorithmic methods did not differ so greatly from ours as this contrast in aims would suggest. Skill was necessary to apply the algorithms, and many successive attempts might be needed to obtain a solution. Nevertheless, the difference in ultimate aims was clear, with the algorithmic methods looking towards a continual reduction in the need for skill, and our own aiming to foster skill and cooperate with it.

The contrast inevitably brings to mind Frederick Winslow Taylor's 'Scientific Management',[1] with its goal of eliminating skill and responsibility in the worker: 'Under our system the workman is told minutely just what he is to do and how he is to do it; and any improvement which he makes upon the orders given to him is fatal to success'. My unease was expressed in the 1975 paper on 'The future of control'[2] and more strongly in a number of later papers[3].

The spirit in which future technology was to be developed had a particular importance at that time. It was clear to those involved in the developing fields of microelectronics, computers, and communications, that very great changes were impending throughout industry and commerce, and in many professional activities such as engineering and medicine. Public awareness came later, but already the outlines of the developments were visible.

A period of rapid change in technology offers great scope for experimenting with alternative routes of development. Once the new techniques have become established, it becomes much more difficult to examine alternatives: the 'entrance fee' for the alternative is the cost of matching the performance of the dominant technology, and it rapidly becomes so high that options which were available become closed off.

For this reason it seemed that the period of 15 to 20 years after 1975 would be one of critical importance. If new technology was developed in a Tayloristic spirit, the working conditions of very many people would be severely damaged. Their scope for initiative and the exercise of skill would be reduced, and work would become more mechanical, more routine, and less human. Because of the power of new technology, this process could be carried further in the traditional domain of Taylorism – production and some office work – and could be extended to new areas, such as design and medical diagnosis, which had been largely unaffected in the past.

In 1979, the generous award of a five-year Senior Fellowship by the Science and Engineering Research Council (SERC) freed me to work on this question, which has two interrelated aspects: 'Why is Taylorism so widespread?' and 'Is it possible to demonstrate a better alternative?'

Why is Taylorism so Widespread?

Explicit support for Taylor's system of 'Scientific Management' is now uncommon, though it can still be found in surprising places. On recently quoting Taylor's dictum, given above, to the eminent Production Manager of a large British company, I was told 'But I agree with that. It's right.'

More generally, Taylorism is seen as an early and crude attempt, which is now outdated, to bring system and order into production engineering. Later developments such as job enlargement, job enrichment, and autonomous groups are stressed, which are ways of alleviating some of the worst features of work organised on Tayloristic lines. Yet if one looks at the actual practice of designers of production systems, the underlying spirit is as strongly Taylorist as ever. The same tendencies can be seen in the development of computerised office systems and CAD systems, though computer scientists are perhaps more generally aware of the problems this brings than are production engineers.

Reactions to this situation tend to be strongly influenced by political persuasions. Condemnation is stronger from the left than from the right, though we can note the pronouncement of Pius XI, 'from the factory dead matter goes out improved, whereas men there are corrupted and degraded'.[4] Braverman[5] argues that Taylorism is an inevitable consequence of capitalism, from which it follows that political change must precede any change in work organisation or in machines. To the objection that political change in Russia and Eastern Europe has led to an entrenchment of Taylorism, he replies that the technology is after all a capitalist technology, adopted by Lenin[6] as an expedient to escape from Russia's desperate and dangerous industrial backwardness, and entrenched by international competition.

This might have been acceptable at an earlier date, but 70 years after the Revolution one would expect to see at least the beginning of something different. The question was certainly raised in Russia, where in the early 1920s there was a vigorous debate[7] on whether a different kind of technology should be sought. Those who thought it should be were defeated, and the question has never been re-opened.

My own views are strongly coloured by contacts with an East European country where new manufacturing technology is under development. Both industrial practice and the aims of research had strong Taylorist tendencies. Attempts to persuade those concerned that a better course was possible met with no success, and I was left with the conclusion that no project such as the one described in this book would have found any support there.

Noble[8] emphasises another but allied causation for Taylorism, namely management's efforts to obtain and extend control over the workforce. Though he stresses the link with capitalism, one can readily extend the argument to any bureaucratic management system. There is the experience of socialist countries to support the extension to them, and my own contacts with large-scale workers' cooperatives, whether the capital is state-owned or worker-owned, indicate that it can be applied there also.

But to say that bureaucracies always attempt to extend their control and limit the initiative of those whom they direct is less than an explanation. Why should they do so, when it involves the loss of the contribution which the ability and initiative and skill of those directed could bring? To explain this, one has to look for a reason which extends more widely than capitalism, seeing that all large industrial enterprises, with negligible exceptions, exhibit the same character wherever they are situated and whether capitalist, socialist or cooperative.

My own explanation, which has been given elsewhere,[9] is that a long tradition in science and technology seeks explanations only in terms of 'cause and effect', so that purpose is excluded. Such a habit of thought ends in a blindness to all purpose except one's own. Then in devising a production system involving people as well as machines, the thought that the purposes of people should contribute to production does not arise. The production system, including its people, is seen as a machine existing only to fulfil the purpose of its originator. This outlook is central to Taylorism.

Such a view illuminates the profound difficulty of achieving the 'worker-less factory' which is the goal of so much present-day effort. A workerless factory would indeed be a machine, and to that extent would fulfil the Tayloristic ambition. But though we can, to a limited extent, incorporate human purposes in machines, our ability to do this is strongly circum-scribed. No machine which we can build, and no computer system, can have the purpose of 'keeping the production system working', in the way that that phrase can be understood and implemented by men and women. Machines do not care whether they work or not, and we do not know how to incorporate the purpose which would make them care. So machines will work well until something occurs which was not envisaged in their design.

In themselves, my own beliefs on this matter can be regarded as external to the project, the aims of which were shared by others with different outlooks and beliefs. As they developed, however, they did play some part in differences of opinion which arose during the development and it is for that reason that they are briefly mentioned here.

An Alternative Technology?

Though the stimulus to my own interest arose from work on the CAD of control systems, this is not a suitable area for a more general study of the possible alternative routes which technological development could follow. First, the subject is abstract and mathematical, and therefore difficult to explain to non-specialists. Secondly, I was personally engaged in the subject: having failed to persuade colleagues on technical grounds that the designer's contribution should be central, I could be accused of shifting the ground to a social argument. Thirdly, it could seem that what was being defended was a particular CAD method, rather than an approach to design and the involve-ment of the designer.

A more suitable example to illustrate a non-Tayloristic development of technology was found in some work which had been done in the Mechanical Engineering Department of UMIST between 1976 and 1980. As I was not involved in this early work, some of the difficulties mentioned above did not arise. The details also were more easily appreciated by those who did not have specialist knowledge of the area.

The work was concerned with the programming of a numerically controlled (NC) lathe; that is, a lathe which in normal working carries out its operations automatically under the control of a program, which has been previously prepared and tested. When such machines were first produced for the USAF in the 1960s, they were programmed[10] away from the shop floor in the planning office, by specialist programmers. After being punched on to paper tape, programs were taken to the machine and tested by skilled setters. A process of modification and re-testing was usually needed, and production was then, in principle, left to an unskilled operator who loaded the machine, started the tape, watched the machine for faults and kept it clear of metal chips. The original job of the operator was thus fragmented into three parts – programming, setting, and operation – in a typically Taylorist development which continued existing trends.

This system was often found to be technically unsatisfactory. In the event of a fault, an unskilled operator could not take corrective action. He could only stop the machine and call for assistance, and the machine would be out of operation until this arrived. The utilisation of the expensive NC machines was therefore found to be low. For this reason, operators were often skilled: but then little use was made of their skill, which was only called upon when a fault occurred. There was clearly a mismatch between the designer's intentions and what was found to be necessary for economic operation.

At the same time, the available electronic technology was changing. Early NC machines used expensive and inflexible special-purpose hardware, which was later replaced by microcomputers, offering increasing power and flexibility. Machines so equipped were distinguished as computer numerically controlled (CNC).

At this point in the development, in the mid-1970s, there was evidently an opportunity to move programming back to the shop floor, to the operator. The microprocessor associated with the machine could offer more-or-less natural facilities for the operator to generate a program, without the need for an artificial programming language. Testing and correction of programs would be facilitated and speeded up. Better use would be made of the skill of the operator, which previously was largely unused. Machine utilisation would be improved, because any necessary changes of program to meet unusual conditions could be made without delay. At the same time, the job of the operator would be re-integrated, by giving him responsibility for programming, setting and operating. One operator would, in suitable circumstances, be able to look after two or more machines.

This opportunity for manual data input (MDI) was taken up by a small number of research groups in different countries – USA, Germany, Hungary and possibly others – and also at UMIST. The UMIST system,[11] completed in 1980, was probably the most advanced of its time. Unlike other systems, it totally abandoned the previous style of programming. It used push-buttons under a dynamic menu, with terms familiar to the operator: 'turn', 'face', 'bore', 'drill', etc. The initial and final diameters were entered in response to questions 'from diameter . . .', and 'to diameter . . .' and similarly for lengths of turned sections and features such as chamfers, fillets and radii. The cutting conditions – speeds, feeds and depths of cut – were calculated by the computer to give improved economy.

It was decided that this system could form the basis for a demonstration of an alternative and better development of technology. It already offered benefits in economy and in operator control which would make it attractive, particularly to small shops handling a wide variety of work. At the same time, it was seen by most production engineers as only an interim solution. The long-term aim, towards which all serious research was directed, was to eliminate human programming entirely by producing a CAD/CAM system (M for manufacturing). A designer, it was envisaged, would generate a three-dimensional description of the part to be made on a CAD system. Then by an entirely automatic process the instructions for the NC machines would be generated from the description held in the computer. Such systems have existed in the aircraft industry for many years, but are at present too expensive for widespread use.

The aim of the proposed demonstration would be to carry on the development of the UMIST system in a way that would maintain its competitive position in the face of this Tayloristic orthodoxy. My belief was, and is, that given equal research effort, a route which makes better use of human ability will always maintain its superiority to a Taylorist route, which rejects this ability. We could not hope to put equal effort into the non-Taylorist research, but a modest effort could very probably prolong the competitive lifetime of the more humanly satisfactory solution.

There was a danger in this particular choice for the demonstration project. This was that the initial superiority of existing MDI systems would ensure them such a commercial and technical success that the project would seem to be unnecessary. This has to some extent happened: MDI systems similar to the UMIST system, and with advanced graphical facilities, are being sold and used on a very wide scale. The result has been (by late 1987) a widespread optimism[12] and a belief that the aims of the project – of a working situation which offers scope for the exercise and development of skill, and for discretion and judgement – have already been met, and this by the normal process of technological development.

My view is that this optimism is misplaced. The underlying aims of most research are unchanged, and as Taylorist as ever. The test of our work will come when MDI systems begin to be seriously attacked by fully automated CAD/CAM systems. We hope that our research will allow MDI to remain competitive, and be further developed, rather than superseded.

The Steering Committee

There is a very extensive literature[13] in the social sciences on job redesign; that is, on taking existing jobs and modifying them to remove some of the worst of their Tayloristic features. A number of techniques are commonly used:

Job rotation, in which workers move around periodically from one task to another

Job enlargement, in which several tasks are combined and given to one worker

Job enrichment, in which some higher-level functions such as inspection are given to workers in addition to their production task

Autonomous groups, where a group of workers is given a long sequence of small tasks, and allowed to divide these and recombine them as they wish among themselves.

These techniques are typically applied to the most fragmented and trivialised tasks, when these are found in practice to be so boring that productivity is reduced and labour turnover is high. As the jobs which are redesigned were produced at an earlier stage by designers, engineers and technologists working on strict Taylorist principles, the procedure has few claims to rationality.

My conviction in 1980 was that if social scientists, with their experience of job redesign, could be incorporated in the design teams, then it should be possible to produce systems which were not only satisfactory from the technological and economic points of view, but which at the same time offered a better kind of work. In line with what was said earlier, the work should offer scope for initiative and for the exercise of existing skills. It should not attempt to preserve skills in an unchanged form, but should provide scope for them to evolve in parallel with an evolving technology.

As this book will show, the aim of integrating technological knowledge with social science proved more difficult than I had assumed. As seen in 1980, however, the aim was to set up an interdisciplinary research team to carry on the development of the UMIST MDI system. Since the guidance of such a team would require skills and experience going beyond my own, I persuaded a number of friends with different backgrounds to form a Steering Committee for the project. It was envisaged that this would meet about four times a year to receive reports on progress from the research team, and to make comments and suggestions for the future direction of research. Individual members of the Steering Committee would also be available to the research team for consultation on problems in their own areas.

The first meeting of the Steering Committee was held on 19 November 1980, after the decisions recorded above concerning the area of research had been taken, but before any application for funding had been made. The initial composition of the Committee was as follows:

Allan Chatterton
Mike Cooley
John Davies
John Fox
Lisl Klein
Howard Rosenbrock

and to these were added subsequently:

Bob Asquith
John Boon, who had attended from the beginning
Harold Palmer

A number of others attended as observers from time to time with the agreement of the Committee, especially:

Håkon Finne

Mammo Muchie

and other visitors and members of staff of UMIST were present from time to time. All members served in a personal capacity. Brief curricula vitae (CVs) of the members (and others) who have contributed to this book are given in Appendix 1 at the back of this volume. At the suggestion of Lisl Klein, meetings from 13 February 1982 onwards were tape-recorded.

As will be seen from the CVs the backgrounds of Committee members covered a wide range, including production engineering, control engineering, trade union activities, management, socio-technical design, and the application (by John Fox) of artificial intelligence (AI) in medicine. It was thought that AI could provide techniques allowing better interaction between operators and a computer. In the event AI was not used, for reasons given in Chapter 5. We were in fact probably six to eight years too early in envisaging the incorporation of AI at the research stage.

The Application for Funding

With the cooperation of the Steering Committee, a research-grant application to SERC was drawn up jointly by John Davies and myself. A copy of the application is given in Appendix 2 at the end of this volume. It envisaged not only the further development of the MDI system for the NC lathe, which then existed in working form at UMIST, but the extension of the same ideas to an elementary flexible manufacturing system (FMS). This also existed at UMIST and consisted of the NC lathe, an NC milling machine, and a robot.

The application asked for three research assistants (RAs), and it may seem that the aim of developing software for even a small FMS in three years with only three RAs was hopelessly optimistic. There was, however, another application by John Davies and Mike Edkins before SERC at the same time, asking for funds to develop integrated software for the FMS. This envisaged two RAs and there were in addition three members of staff of the Mechanical Engineering Department involved. Thus if both applications had succeeded, there would have been, besides myself, three staff members, five RAs, and a number of Ph.D. students involved in the research.

In the event, the application given in Appendix 2 succeeded, resulting in a grant dating from 1 January 1982. The other application unfortunately failed, and as a consequence little further work was done on the FMS by the Mechanical Engineering Department. Also, as explained below, the integration of engineering and social science gave more problems than were envisaged at first. Research was therefore concentrated on the MDI system for the lathe. Nevertheless, it is believed that much of what was learned in this narrower area has more general applications.

Steering Committee Discussions

It will be noticed that there was a long delay between the application for funding and the grant (1 January 1982). There was a further delay in recruiting the RAs:

Martin Corbett (social science), 5 July 1982

Roger Holden (computer science), 4 October 1982

Paul Kidd (control engineering), 1 May 1983

It was therefore only in mid-1983 that effective research began on the combined technical and social design.

During this time the Steering Committee continued to meet. There was some difficulty in maintaining enthusiasm during the long period while funding was in doubt. There was also a lack of specific research on which to comment from late 1980 to some time late in 1982. During this period the Steering Committee raised and discussed a number of basic issues.

One of these became known as the 'hand and brain' issue. Modern technology continually replaces manual tasks by the task of overseeing and controlling an automatic device. In the eighteenth century, the turner held the tool and moved it manually on a tool rest. To produce a true cylinder was a matter of great difficulty and skill, and screw-threads were still more difficult to produce accurately. In the early nineteenth century the slide-rest was introduced, which held the tool and moved it in a straight line. The operator moved the slide-rest by turning a handle, but screws were formed by, in effect, turning this handle through gearing. 'Power-feed' later eliminated the need for the operator to turn the handle for most operations. Numerical control took all manual functions away from the operator.

The questions raised were whether this development was desirable (some members feeling strongly that it was not) and whether an aim of the project should be to retain or re-introduce some manual element into the operator's job. The issue was debated at length within the Committee, and with social scientists outside the Committee. No final conclusion was reached, though something is said about the question in later chapters.

My own view is sympathetic to the retention of manual skill. I have a workshop with a lathe, milling machine, drill, etc. which are all manually operated, and I enjoy working with them. On the other hand, if I were engaged in manual work all day and every day I am not sure that my attitude would be the same. But the major difficulty is that I can see no way of protecting the element of skill and control in work with a lathe, in competition with alternatives, in a way that resists the mechanisation of manual work, whereas I can see such a route if mechanisation is accepted.

Another similar discussion concerned 'digital versus analogue' representations. Is there something inherently more suitable for human beings in analogue devices rather than digital devices? This also aroused strong feelings, though I do not have a fixed opinion on the subject myself. In fact, as will be seen later, we found an analogue presentation of the constraints on cutting conditions (by means of a bar chart, e.g. Fig. 4.1) to be much better than a digital or a verbal presentation.

Methodology

A particularly important meeting of the Steering Committee was the one held on 12 July 1982. By invitation, Ken Grainger of Coventry Workshop attended this meeting. He had previously visited UMIST to discuss the research that we were proposing to do, and had expressed the strong view that we ought to involve workers directly in the design process. I invited him to put his view to the next meeting of the Steering Committee, and a verbatim transcript of the discussion which took place is given in Appendix 3 at the end of this volume.

Several distinct strands can be discerned in the discussion. First, there was resistance to the view that the work should be explicitly aimed at trade union objectives. There are projects, such as the UTOPIA project mentioned in Chapter 7, which have been set up with direct trade union involvement, but it was felt that this was inappropriate in the UMIST project, just as control by a management organisation would have been.

Secondly, practical difficulties were seen in involving workers in the project. In a long-term research, the individual workers who will use it cannot be identified. The population from which they will be drawn can be defined by the research team, but there is no guarantee that this definition will be adhered to when the equipment is brought into use. The definition itself poses difficulties – if the aim is to re-integrate the operator's job by giving him responsibility for programming, one excludes parts-programmers from the user population. As programmers and operators belong in Britain to different unions, this causes further difficulty. There are also problems in arranging for workers to participate, as funding bodies do not normally cover such costs, and industry may be unwilling or unable to do so.

Besides these practical difficulties, which did not question the principle of worker participation, further difficulties were raised from the social science side which did contest the principle. Workers associated with the project might have only a token role, leaving the design process unaltered. They might come to be identified with the project by other workers, and accused of adopting a patronising rather than a representative role. They might have difficulty in contributing effectively at an abstract level in design, whereas if they were presented with concrete alternatives their opinions would be much firmer.

The discussion on 12 July 1982 largely determined the methodology which was adopted in the project. This was to expose the design to comment and criticism by workers at various stages in the development process. The software was developed on a terminal connected to a large Prime computer away from the lathe, so the demonstrations had the nature of a simulation. A good sample of workers would probably be 100, but this was always unrealistic in the context of the project. We aimed at 12, and never had more than 5.

While the earlier MDI system remained in operation, visiting operators were first given an opportunity to program with its aid, and to turn a part using its facilities. They were then introduced to the simulation. Later, the MDI system suffered from lack of maintenance, it became unusable, and operators had to use the simulation alone. To those unfamiliar with NC machines, this presented an artificial situation which was difficult to relate to the ultimate shop-floor experience.

An extensive account of these demonstrations to operators is given by Martin Corbett in Chapter 3. My own experience has led me to be strongly doubtful about their value. First, what one wants from operators is a view on how they would regard the system after a considerable experience in a real working situation. To bring them into an unfamiliar environment for one day, and to show them a half-developed system in simulation, is most unlikely to produce any approximation to their long-term views.

Secondly, the design process does not generally produce a number of clear-cut alternatives at an intermediate stage. One has rather a multitude of small decisions, guided as in Geoffrey Vickers' quotation[14] by 'the elimination of "misfit". The designer approaches his task with a number of tacit criteria, which appear only when some specific design is found to be inconsistent with one of them. The norm is known only negatively, when it is infringed'. In other words, the process is driven by the designer's discontent, and once he has destroyed this he has destroyed at the same time his motivation to produce valid alternatives. What is needed is the generation of discontent, not only with designs which are technically unsatisfactory but also with those which do not satisfy the social criteria. This may need to be done by communication between different members of the design team, rather than by an individual.

Thirdly, it encourages a tendency to offer options to the operator: 'would you like to see this option included?' There is an inclination to answer 'yes' to this question in relation to any single option, and one can then accumulate a large number of them. Each incurs a penalty in development effort, and also makes the system more complicated and difficult to understand and use. Having put an option in, one has shifted the burden of proof: one is no longer looking for reasons to put it in but for reasons to take it out. The methodology has an inherent tendency towards complication.

A great many small decisions were the subject of debate between the representative of social science in the research team (Martin Corbett) and the representatives of technology. Rather than dissipate energy on so many, one representative decision was taken as an example for deeper consideration. This was the 'blank table option' which is discussed in Chapters 3, 4 and 6. It was not chosen as having greater importance than any other, but rather the opposite – it was chosen as typical of a multitude of others.

The 'blank table debate' is adequately dealt with elsewhere, but in relation to methodology it is worth noting that during the course of the debate, five operators were brought in to assess the then state of development. The blank table option was not available, but it was described to operators and they were asked if they would like to have it. None of them did, but this was not taken as settling the debate. That fact strongly reinforced my own doubts about the methodology we were using.

My own present view is that I would like to see operators involved in the design process, but doubt whether it is possible, with our present knowledge, in a project such as this. It has been done effectively[15] when the technology is simple and easily appreciated, and cheap and quick to simulate. It has also been done[16] when the question at issue is which of a number of existing systems to select. It has not been done, so far as I know, in a project which is long term, which aims at an ill-defined population of users, and which is technically complicated.

Empirical testing, I believe, can be done only when a complete system has been developed, and has been in use for several months in a real production environment. The methods of social science are well adapted and well developed for assessing such a situation.

It may seem that in saying this I am relegating social scientists to their traditional role of retrospective study, and excluding them from the design process. That is not my intention. One of the lessons of the project is that retrospective studies, as they have been carried out in the past, are of limited value to the social scientist as part of a design team. What I envisage is a design team with a continuing brief. First it might develop an MDI system for a lathe, and put a prototype into industrial use, recognising that modifications would most likely be necessary.

These modifications would be worked out on the basis of reports of shop-floor experience, obtained by a social scientist who was a part of the design team. At the same time, the team would be proceeding with the design of further systems, say for MDI of milling or drilling. What had been learned from shop-floor experience with the lathe would also, at least partly, be applicable to these later developments. The beginnings of such a process (though only the beginnings) will be possible in the ESPRIT project (Chapter 11).

A social scientist in such a situation would be able to speak with much greater authority about workers' opinions. He would also, by virtue of his membership of the design team, have a different outlook on what was needed by designers from his investigations into the reactions of workers. The need, for example, for some principles to guide the designer on the human implications of decisions, as described in Chapter 3, might well assume a greater significance than it would if he were divorced from the design team.

The disappointing thing about this conclusion, if it is valid, is that guidance from users would be most valuable in the early stages when the design is most malleable, but it is then that reliable guidance is most difficult to obtain. The problem deserves serious study: possibly better ways of simulating or presenting the tentative designs, or better ways of explaining them to users, can be found.

The Aims of the Project

As will have been gathered, the aims of the research were never so firmly settled that they were regarded as beyond question. Debate on this question continued throughout the life of the project. One way which I have found useful for describing some of the alternative views is the following, which is an elaboration of something used many times in discussion.

In Fig. 1.1, P represents the present state of a particular technological area, such as MDI for lathes – or indeed it might represent the whole of the present technological situation. In P, there is a sun S, which represents the sum of all

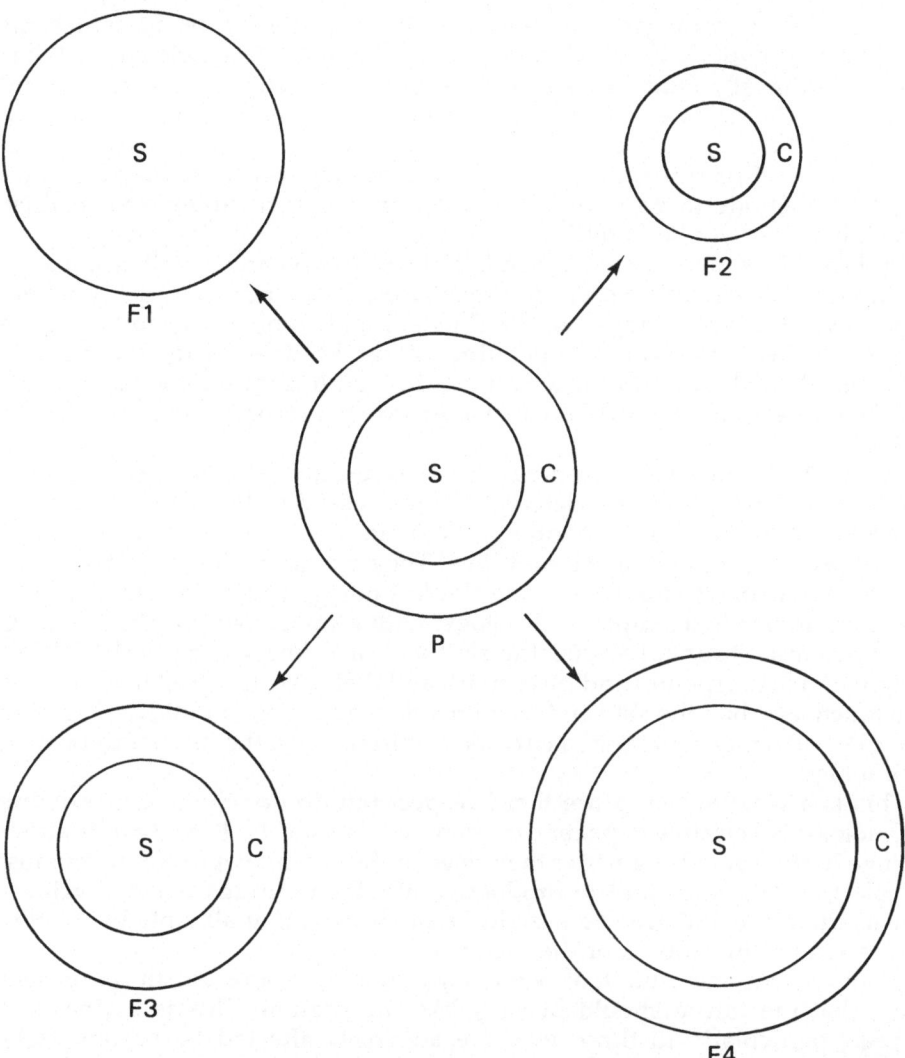

Fig. 1.1. Four different views of the evolution of technology. In each of the diagrams S represents the 'sun' of explicit knowledge, while C represents a 'corona' of skill and tacit knowledge without which the explicit knowledge cannot be used. F1 is the Taylorist view of the future; F2 rejects much of our present technology; F3 sees the future as no different from the present. In F4 a future is envisaged in which the explicit knowledge has increased, but scope and opportunity have been given for a corresponding development of skill and tacit knowledge.

explicit – or, if one likes, 'scientific' – knowledge about the area. This is surrounded by a corona C, which represents all the skill and tacit knowledge of the area, without which the explicit knowledge remains sterile and unusable. No-one can learn to play golf simply by reading books about it. In the same way, no amount of theoretical knowledge about metal-cutting machines will allow one to use a lathe effectively, without the experience and skill which render the book-learning fruitful.

Around P are arranged four possible views of the future. F1 represents the Tayloristic ideal, in which all of the skill and tacit knowledge in P has been scientifically studied and made explicit. The sun has expanded and swallowed the corona. The technology now, in principle, has no need of human assistance, and can be entirely automatic. If for economic reasons people are still retained, they can be told minutely just what they are to do and how they are to do it, and they exist only as substitutes for machines which have not yet been built.

In fact F1 is unattainable. Some human intervention will always be required to meet unforeseen circumstances. Machines will malfunction, power will fail, workshops will be flooded, and human ingenuity will be needed to keep the system operating. What F1 does is to devalue this necessary knowledge and make it more difficult to acquire and to exercise. The Japanese 'quality circles' are to a large extent a recognition of the defects of F1.

F2, on the other hand, represents an opposite extreme. Because of a sensitive awareness of the defects of our present technology, and of the directions in which it is developing, there are some who wish to reject it,[17] and speak approvingly of the Luddites. They would wish to reject some of our present explicit knowledge, and the technology in which it is incorporated, and return to a simpler technology, with a smaller sun S. The corona C in F2 has also shrunk, because the skill and tacit knowledge of the wheelwright and the carpenter and blacksmith and their like, taken all together, are still much less than the skill and tacit knowledge needed to build and operate aeroplanes, to make integrated circuits, and to carry on the rest of our present technology.

This is a view of the future that I respect but do not share. Our existing technology is certainly repellent in many of its aspects. But I would rather change it into something better than reject it. Rejection would condemn most people to a laborious and unproductive life. Some things which I value – music, literature and science – could be carried on, if at all, only by an élite supported by the labours of the majority.

F3 represents the view that technology should be arrested in its present state, that the future should be very like the present. This is a view that appeals particularly to those who are adversely affected by current technological change, and is reflected in many trade union views. Again I have sympathy with it, but cannot share it. Certainly the development of technology should not be allowed to make existing skills and experience suddenly obsolete. Certainly also the evil of unemployment should not be tolerated. But to achieve these things by arresting technological change would require a world-wide agreement which it is unrealistic to expect. It would also reject, together with its damaging effects, the positive advantages which technological change could bring, not least in correcting the evils of present technology.

F4 represents a future which I believe to be possible and would like to see. The sun of explicit knowledge has expanded as much or more than in F1. The corona of tacit knowledge and skill has expanded in proportion, and this has been achieved by design, not by accident. The development of technology has been guided in such a way that at each stage the sum of existing skills is relevant and useful, but these are not regarded as static. The technology gives room and encouragement for existing skills to change with time into the new skills which the changing technology requires. Productivity increases as much or more than in F1, but unemployment is avoided by a reduction in the working week, by study leave, by earlier retirement, by better housing and education, or by a multitude of other ways in which the benefits of increased productivity can be distributed. Some part of the increased productivity could also be used to eliminate those unattractive features that mar our present technology, such as pollution and ecological damage.

Throughout the project there was a continuing debate around this issue. It seems that many social scientists cannot accept F4, or do not believe in its possibility. For example, the 'blank table debate' reported in Chapters 3, 4 and 6 started from a minor technological decision, but developed into a questioning of F4 – equating it indeed to F1 – and an apparent preference for F3.

References

1. Frederick Winslow Taylor, On the art of cutting metals, 1906, 3rd edn (undated) p. 55 (American Society of Mechanical Engineers).
2. H.H. Rosenbrock, The future of control, Plenary address, Sixth IFAC congress, 1975; Automatica, 1977, vol. 13, pp. 389–392 (Pergamon Press).
3. H.H. Rosenbrock, Engineers and the work that people do, IEEE Control Magazine, 1981, vol. 1, pp. 4–8; Robots and people, Measurement and Control, 1982, vol. 15, pp. 105–112; Designing automated systems – need skill be lost?, BAAS annual conference, section X, 1983, Brighton.
4. Pius XI, Quadragesimo anno, 1931, quoted in Work and the future, 1979, p. 20 (CIO Publishing).
5. Harry Braverman, Labor and monopoly capital, 1974 (Monthly Review Press).
6. See V.I. Lenin, Raising the productivity of labour, Collected works, vol. 27 (Feb – July 1918) pp. 258–259 (Lawrence and Wishart, 1965).
7. Mammo Muchie, Coupling Taylorism with socialist doctrine – dilemmas and debates in the early twenties. Personal communication.
8. David Noble, America by design, 1977 (Alfred A. Knopf).
9. H.H. Rosenbrock, Machines with a purpose, to be published.
10. See ref. 8.
11. J. Boon, L. Satine, S. Hinduja and G. Vale, Back to operator control?, Numerical Engineering, April 1980, vol. 1(2), pp. 27–29.
12. See, for example, Bryn Jones, Destruction or redistribution of engineering skills? The case of numerical control, in Stephen Wood (editor) The degradation of work?, 1982, pp. 179–200 (Hutchinson).
13. For a conspectus, see Richard I. Drake and Peter J. Smith, Behavioural science in industry, 1973 (McGraw-Hill).
14. Geoffrey Vickers, in Judith Wechster (editor), On aesthetics in science, 1978, p. 148 (MIT Press).
15. Lisl Klein, A social scientist in industry, 1976 (Gower Press).
16. Enid Mumford, Designing secretaries, 1983 (Manchester Business School).
17. See, for example, David Noble, Present tense technology, parts 1, 2, 3, Democracy, 1983, spring, pp. 8–24, summer, pp. 70–82, fall, pp. 71–93.

Chapter 2

The Technical Problem

Howard Rosenbrock

Introduction

This chapter has two aims. The first is to describe the technological background against which the project was conceived. The second is to outline a collection of problems which were debated throughout its progress, both in the Steering Committee and in the research team. These problems were all concerned in one way or another with the relation between people and machines in the production process. The issues raised therefore lie partly in technology, partly in economics, partly in sociology and other areas of social science, but also very largely in the uncharted territory which lies between these disciplines.

The method adopted will be to describe the problems. This is inevitably done from a personal viewpoint, but without insisting too much on personal preferences in their solution. A wide range of questions will therefore be raised. Only at the end of the chapter are strongly personal views on their solution suggested.

Some History

The use of digital electronics to control machine tools was a natural development from an earlier stage of automation. Let us take the lathe as an example, because it is one of the most basic, and one of the first machine tools to be developed.

By the eighteenth century, after several millennia of use,[1] the lathe had developed to the form shown in Fig. 2.1. Motive power was provided manually by turning a large handwheel, and Dostoevsky[2] describes how he and another convict would labour at this work, under almost unchanged conditions, in the mid-nineteenth century: 'great efforts were necessary to

Fig. 2.1. The lathe as it existed in the eighteenth century.
A belt connects the large, hand-driven wheel with the lathe. Turning is done by a hand-held tool supported on a rest. (From Diderot and d'Alembert, Encyclopédie ou Dictionnaire Raisonné, Tome 27, 1771, Article Tourneur, Plate 1, after p. 19.)

make it go round, above all when the workmen . . . used to make the balustrade of a staircase or the foot of a large table, which required almost the whole trunk. No one man could have done the work alone.' Compared with other prison tasks the work interested him: 'As for me, I found the exercise most salutary'.

The large wheel was connected by a belt to a smaller one mounted on a spindle, driving it at a higher speed. Work to be machined – wood or metal – was mounted between two pointed 'centres' and driven by the spindle. A flat rest, close to the work and at a convenient height, was arranged to support the tool. Then the turner held the tool by a long handle, and pressed it against the rotating workpiece. Shavings of wood or metal were removed until the workpiece was round in section and had the required shape in its length. The turner's work, says Rolt, 'called for considerable muscular effort as well as skill'.[3]

By 1800 in England steam power was being used to drive the lathes, and the increased power was making it difficult for the turner to control the tool. The demands for accuracy made by the new machines for spinning, and by steam engines themselves, could not easily be met. To turn a true cylinder of constant diameter – say for a piston rod – was a matter of great care and skill, taking much time. Screw threads were also very difficult to produce. A 'comb tool' with many teeth, spaced according to the thread to be produced, was brought up to the rotating work. The tool was moved along at exactly the right speed to allow the second tooth to enter the groove made by the first, and so on. Once the thread had been started, the tool would follow the existing grooves, and gradually deepen them. It was a slow process, and the accuracy achieved was not high.

To meet these difficulties, the slide rest and lead screw were developed[4] by Maudsley and others from 1800 onwards. Instead of being freely supported on a flat rest, the tool was clamped in a tool holder. This was attached to two sliding metal frames. One moved the tool along a path parallel to the axis around which the work rotated. The other, the cross-slide, could move the tool along an axis at right-angles to the first, towards or away from the work. Each slide was driven by a screw.

The forces at the tool were now supported by the slides, and they could be much greater than before. To produce an accurate cylinder was straight-forward. The turner set the tool at the right distance from the axis by means of the cross-slide. Then by means of a handwheel attached to the screw which moved the other slide, he could traverse the tool along the work, parallel to its axis. The accuracy of the cylinder produced did not now depend entirely on the skill of the worker, but was largely built into the machine, through the accuracy of its slides. A before-and-after sketch (Fig. 2.2) given by Nasmyth[5] in 1841 illustrates the development with some elegance and wit.

The evenness with which the operator turned the handwheel to move the slide did affect accuracy, but this too could be made automatic in a number of

Fig. 2.2. The evolution of the lathe in the early nineteenth century.
The turner on the left is working in the same way as in Fig. 2.1, though his lathe is now power-driven. On the right, the new generation is using a slide-rest. (From James Nasmyth, in Robertson Buchanan (editor), Practical essays on mill work and other machinery, 1841, p. 396 (John Weale).)

ways. For example, by means of gears, the screw moving the first slide could be driven from the lathe spindle, causing the tool to feed forward a definite amount for each revolution of the work. The amount of this feed could be changed by altering the gears, which became known appropriately as change wheels. The operator now no longer turned the handwheel to move the tools, but simply engaged or disengaged the mechanical drive. Screw threads were easily produced by using a large feed equal to the distance between successive threads. By a later extension of the same idea, the cross-slide could also be driven automatically for machining the flat face of a disc such as a flywheel.

This arrangement, known as 'Maudsley's go-cart' to sceptical turners, has obvious advantages for much routine work, but there were also losses. Cylinders and discs and screw threads were made quickly and easily. Conical parts could be turned by arranging for the axis of rotation of the work to be no longer parallel to the slide. But other shapes – spheres and free-hand curves – became more difficult.

Whereas with a hand tool and rest the turner could produce an arbitrary shape by eye, or to a template, this was no longer practicable with the slide rest. The turner could revert to hand tools, or a special 'form tool' could be made with the required shape of the work ground into its cutting edge, and this tool could be driven into the work by the cross-slide. Or, as was done later in the nineteenth century, special and more complicated 'copying lathes' could be used to move the tool automatically so that it followed a path defined by a master template. In fact, for some kinds of work such as ornamental brass work or wooden chair legs, simple lathes without slide rests were used well into the twentieth century, and the skill of hand-turning persisted.

This highly simplified account is a retrospective interpretation of a complicated historical development.[1] It has been given for the sake of those unfamiliar with machine tools and their development, and because of the light it throws on some of the discussions in the Steering Committee. The technical gains and losses which arose from the introduction of the slide rest have been mentioned briefly before, but there were other repercussions.

A skilled turner in the early nineteenth century, who had acquired his skill with hand tools over many decades could certainly see 'Maudsley's go-cart' as an attack upon this skill. Some kinds of work, such as turning long, true cylindrical parts in steel, had required the continuous application of skilled handwork over many hours. With the slide rest the work could be set up in the lathe, and could then proceed automatically for long periods. The tool would start at one end of the work, and traverse automatically along it, taking perhaps twenty minutes to reach the other end. Then the tool would be withdrawn by the turner and the slide returned to the starting point. The tool would be moved in towards the work, and the drive engaged for a new cut. Apart from stops to exchange worn tools, or check the size of the work, turning went on in this way continuously and automatically. A previous skill in using hand tools was no longer needed for this kind of work.

At the same time, new skills did become necessary. In order to cut a screw thread, the correct set of change wheels had to be selected from the limited number available. Standard threads would have standard set-ups, but the ability to select the appropriate change wheels for a required thread became a necessary skill in the turner. The elementary mathematics needed, in the days before compulsory schooling, could be unfamiliar, and such new requirements go some way to explain the rapid growth of Mechanics' Institutes, and the determination with which instruction was sought.

Now the early nineteenth century was dazzled by the idea of progress and by its visible manifestations. Though there was dissent, the prevailing attitude was strongly opposed to any questioning of the changes which were taking place. The late twentieth century has a different perspective, which underlay the 'hand and brain' issue mentioned in Chapter 1. How should we now evaluate the changes which took place in work, and in the skills it required, when hand-held tools were superseded by the slide rest?

'Striking the thread' with a comb tool meant moving it into contact with the work in precisely the right way to create a helix which could be followed by subsequent teeth. It was a psychomotor skill acquired only by long practice. Plain turning with the hand tool was another psychomotor skill. Michael Polanyi[6] remarks that when we use a tool, for example a hammer, we do not feel the reaction between the handle and our hand. What we 'feel' is the reaction between the hammer head and the nail, which by a hidden, tacit process our body generates from the forces at our hand. In the same way the turner using hand tools 'feels' the cutting action of the tool on the metal.

These very direct skills were largely displaced by the development of the slide rest. In their place arose new skills which were entirely mental – calculations of change wheels, for example. Some psychomotor skills did remain, in an attenuated form, because when the automatic feed was disengaged, and metal was cut by turning the handwheels which moved the slides, it was still possible to 'feel' what was happening at the tool, though less directly and accurately. At the same time, the physical work of holding and moving the tool in turning to take heavy cuts became unnecessary. Work was done more quickly, and it was done more cheaply. Labour productivity increased, but less rapidly than production.

The recent development of numerical control, described below, carries the same tendencies much further. Labour productivity is now rising faster than production, which creates unemployment in the absence of any better ways than we have of distributing the wealth produced. The last remnants of psychomotor skills are also displaced, and the machining of metal becomes intellectualised, as an interaction with a computer.

The 'hand and brain' issue, which is mentioned in a number of later chapters, concerned this final loss of psychomotor skills. Is it tolerable, or desirable? If not, can an alternative development of technology preserve those skills, for example by retaining handwheels? Any such discussion must pay regard also to the earlier development. For example, if it is important to retain handwheels, would it not be equally important to revive the skills of turning with hand tools which were lost at the time when handwheels were introduced? Any coherent view, if it does not simply reject (or simply accept) all developments which substitute mental for physical skills, must possess some criterion to distinguish those which are acceptable from those which are not. There is also the difficulty that an argument for retaining psycho-motor skills may be pressed, not on its inherent merits, but as a way of retarding the increase of labour productivity, and the manifest evil of unemployment which in present conditions results from it.

Numerical Control

During the nineteenth century, and the first half of the twentieth, there was continual progress in making lathes more automatic in action. This was done by mechanical means such as cams, which were specially made for each part. Making the cams required much care and time, but the machines were very suitable for mass production. A cam automatic could produce all the tool movements needed to make a turned part, as well as feeding forward the bar from which the parts were made. It would be quick and accurate in operation once set up, but changing from the production of one part to another would be slow, even if the cams were already available.

When the technology of digital electronics was developed after the Second World War, there was an obvious possibility of using digital control for machine tools. A cam, for example, will produce a particular displacement of the tool for each angle through which it is turned. A table of these displacements, at close intervals, can be stored digitally and is equivalent to the cam. In principle, a new table can be inserted in place of the old one very quickly by reading it from a storage medium, which originally was punched paper tape.

In fact lathes were not the first machine tools to be automated in this way. The easiest machines to control digitally were jig borers, which are used to bore holes in a block of metal at very precisely defined positions. The boring head is set to a desired position, a hole is bored, the head is moved to a new position, and so on. To define the positions digitally, leaving the boring to be controlled mechanically, was a relatively easy task.

It is considerably more difficult to move a tool in a continuous path at a desired rate. For example, if an ordinary lathe tool is to be used to turn a spherical surface in a lathe, the tool has to be moved around part of a circle by the simultaneous movement of two slides. A digitally produced circle is in fact a series of very short straight lines, so that the shape produced is imperceptibly different from the one desired. A great deal of information is needed to define these straight lines, and severe demands are made on the digital circuitry and on the automatic drives of the slides.

The first solution to this technical problem of continuous-path control was obtained at MIT in a research programme financed by the USAF from about 1950 to 1955. The machines this time were intended for milling aircraft wings, and the development has been described in detail by Noble.[7]

Early aircraft were made with a wooden frame covered by fabric. This was later replaced by a skin of light alloy, riveted to a light alloy frame as in modern civil aircraft. Demands on military aircraft by 1948 had become so severe that it was proposed to make wings in which frame and skin were a single piece of metal. In the ultimate implementation of this idea, a large slab of aluminium alloy was machined by a rotating cutter. One surface of the slab was brought to the shape of a part of the external wing surface. On the other side, a multitude of pockets were cut, the bottom of each pocket being the inside surface of the wing. The walls between the pockets formed the strengthening ribs which carried the load on the wing. An upper and a lower section, fixed together, gave the complete form of a part of the wing, and several such parts made up the whole wing. In the process of manufacture nearly all the metal in the original slabs was machined away, and for each slab machining would take many days.

To do this work by conventional methods appeared excessively difficult, even with copy-milling machines, which were able to follow the shape of a template. One problem was the time taken. A second problem was the likelihood of error. One mistake at the beginning of the process would scrap the slab of alloy. A mistake near the end of the process would incur the extra cost of all the work that had already been done, and would also delay production. For these reasons the USAF was prepared to support an expensive research programme in order to produce a digitally controlled milling machine. Once this was programmed, and the program proved, wings could, in principle, be produced automatically, and they would all be identical and perfect. Such, at least, are the kind of hopes which inspired the research.

The difficulties and failures and successes of the research can be followed in Noble's account.[7] Digitally controlled machines were achieved, and were produced commercially. At first the digital electronic controllers were specially designed and produced for their task, and the machines were described as numerically controlled (NC). Later it became more economical to use generally available computer chips, and design became largely a matter of programming. The machines were then said to be computer numerically controlled (CNC). Linking machines to a central computer, so that programs could be sent electronically to their controllers, produced direct numerical control (DNC). At the same time there was a development of computer-aided design (CAD), and means were developed by which the design stored in a CAD system could be translated automatically, by a computer, into the program which would cause the appropriate machine to make the part.

All of this existed in a more or less complete form in the aircraft industry by 1970, but it was not widely used even there for general machining, and diffusion into industry at large was slow. It was expensive, required large computers, and imposed a very rigid organisation. Parts programs were usually not produced automatically, but by parts programmers in the planning office, separated from the machines and using the artificial language APT. Skilled setters proved out the programs, and operating the machines for production was regarded, at least initially, as unskilled. Poor machine utilisation might in fact lead to the re-introduction of skilled operators, whose skill was then under-used, being called upon only when things went wrong.

This organisation reflected the existing Tayloristic subdivision of work, and embedded it in the technical system. Not only was the opportunity to create the program for a part denied to the machinist, but the means of doing it were not provided. Consequently abnormalities arising at the machines, and needing corrective action at the machines, could be dealt with only by an organisational loop involving operator, foreman, production engineer, programmer and setter. An early answer to this problem was 'manual data input' (MDI) which allowed the operator, somewhat laboriously, to change the detailed instructions of a program, line by line, at the machine.

While programming had to be done by writing code in APT, it seemed inevitable that it should be done away from the machines. But as computing power became cheaper, the situation changed. A CNC machine had a computer associated with it for control purposes, and if suitable instructions could be provided, it could be used also to create the programs which were to be run. The instructions need not take the form of a programming language, but could be in terms familiar to the machinist. They could be stored, and recovered for further batches of parts if needed.

This was the background to the development, at UMIST[8] and elsewhere, of programmable controllers, the UMIST development being completed by 1980, before the beginning of the project described in this book. The motivation was not, in this earlier phase, explicitly 'human-centred', though experience with industrial production by some of those concerned pre-disposed them strongly towards developments making better use of human initiative and skill. Rather, the emphasis was on increasing the flexibility and responsiveness of production.

As explained in Chapter 1, this previous development at UMIST formed the background to the 'human-centred' project. For technological reasons it was becoming easy, and was seen to be desirable, to re-integrate the job of the machinist. Programming had been separated from operation, and re-moved from the shop floor. It could now be brought back, so that the machinist could take responsibility for the whole task of machining a component, from definition to execution.

This would be particularly attractive in a jobbing shop, where batches were small, and new jobs were continually coming in, so that programming was a significant part of the work. Its attraction might extend into situations where batches were larger, but there was still a significant amount of new work. In mass production, where only a fixed range of parts was produced, it would offer little benefit in re-integrating work. Once all the parts had been programmed, even if they were produced batch by batch, the situation would be little different whether the programs had been created on the shop floor or in a programming office. Though the idea of a 'human-centred' technology was felt to be as applicable to mass production as to the jobbing shop, the programmable controller was not felt to be the appropriate device for implementing it there.

Record–Playback

To complete the background to the project, it has to be mentioned that an alternative way of programming existed even before the beginning of the USAF/MIT project. This was a scheme by which a machinist would make the first part by conventional techniques. As he did so, all the machine move-ments would be recorded automatically. Then subsequent parts could be machined by replaying the recorded moves. This possibility was a persistent theme in Steering Committee discussions, and some members strongly favoured it.

This early system was termed 'record–playback', and the same term is used in the SERC application (Appendix 2) to describe the system proposed, and largely followed, in the project. The latter has one essential feature of the early systems, namely that the program is to be created by the machinist at the machine, and immediate feedback is obtained by running the program, either operation by operation, or as a whole. On the other hand, the early proposals envisaged the first part being made by conventional means, rather than by interacting with a computer as in the project. It was this difference which caused discussion. In what follows, to avoid confusion, 'record–playback' will be used to mean the creation of programs by the use of conventional machines equipped with digital means for recording their movements.

Noble[9] suggests that record–playback control was rejected in favour of parts-programming for the sake of management control over production. In particular, military aircraft production was strongly cyclical, and the workforce was expanded and contracted to match it. If production knowledge resided in the workforce, it would be carried away with them in a phase of contraction, and would be difficult to recover when production expanded again.

There is no doubt a management pressure to appropriate and retain production knowledge in the organisation, but the application of this to record–playback is not obvious. Once a program has been created by record–playback, it is as available to management as if it had been written by parts programmers. And to retain a small nucleus of skilled machinists to create new programs would be as easy as to retain a nucleus of parts programmers.

A more direct explanation, to my mind, can be found in attitudes that are deeply embedded in the scientific and technological culture, which has great difficulty in acknowledging and accepting human purpose, and the skill which represents a purpose accomplished in work. To the engineer and technological research worker, a system which relies upon the existence of human skill, and particularly manual skill, will appear defective and incomplete. Only knowledge which is explicit and definable is accepted, knowledge which can be embodied in text-books, transmitted to a new generation, and used as a basis for further advance. This attitude, and the culture which embodies it, have been undeniably successful both in science and in technology.

The same attitude is at the root of Taylorism: 'Under the old type of management success depends almost entirely upon getting the "initiative" of the workmen . . . Under scientific management . . . the managers assume new burdens . . . for instance, the burden of gathering together all of the traditional knowledge . . . and then of classifying, tabulating, and reducing this knowledge to rules, laws, and formulae . . .'.[10] One of the major problems, in a project which rejects Taylorism and seeks to find a better role for people in the production process, is therefore to overcome the difficulties which this attitude brings. The question here is whether record–playback is an appropriate means to this end.

This question can be approached in several ways. Technologically, it presents a number of difficulties. First, in the context of the USAF/MIT project, it does little to ease the task of the machinist who creates the program by making the first part. The cost of errors can be reduced by making this 'first part' of plastic, rather than metal. But if a mistake occurs, the machinist will have to begin again, unless means are provided for editing the program. Such means can undoubtedly be provided, though not so readily as when the parts program is created in the usual way as a computer program.

Secondly, there is a more important difficulty in the context of the UMIST project. As explained on pp. 19–20, the development of the slide rest around 1800 forced a change in the way that machining was carried out. With a hand rest, the tool could be moved free-hand by the turner to produce arbitrary contours. With the slide rest, the production of arbitrary contours on an unspecialised lathe demanded a form tool. With numerical control, the ability to move the tool along an arbitrary path is regained, and form tools (which impose a number of restrictions) are no longer needed.

If we were moving from hand turning to numerical control, record–playback would not pose this conceptual difficulty, though it would pose great problems in recording. If we wish to move from modern conventional lathes, with their slide rests, handwheels, and automatic feed, the problem is severe. A turner using this kind of machine is unable to manipulate the tool in a way that will produce the arbitrary shapes which could earlier be produced (though inaccurately) by hand-turning and can now be produced by NC lathes. Record–playback will therefore put great obstacles in the way of using a numerically controlled lathe to its full extent.

A type of hybrid programming has been suggested by Gossard[11] which might overcome this problem. The program is created through a computer interface, rather than by using a lathe, but the interface resembles to some degree the controls of a conventional machine. It is equipped with hand-wheels which simulate those driving a slide rest, but which are connected instead to digitisers. As each wheel is turned, it operates a digital electronic display, which indicates the position of the imaginary tool with high accuracy. At the same time, a visual display indicates the tool position in relation to the work, and can show how metal will be removed.

Because the turner is interacting with a computer, facilities can be provided which are not available on the conventional lathe. For example, the centre and radius of a circular arc can be put in through the handwheels, and the computer can generate instructions to drive the tool along this arc. Arbitrary curves can be entered as a series of points, through which the computer will generate a smooth curve.

What have been described so far are technical difficulties and possibilities. A different kind of question is whether record–playback gives a more desirable working situation. If the program is created at a real conventional lathe, the psychomotor skills, so far as they still exist in the operation of handwheels, are retained. It is then a question how important this is, which returns us to the hand-and-brain issue. If programming is done as in Gossard's method, psychomotor skills are not retained, because the hand-wheels are connected only to digitisers and give no feedback to indicate the forces at the tool. The wheels are simply an input device, equivalent to the keys which could be used to replace them.

Economically, a problem with record–playback is that if a real lathe is to be used for recording, it must be equipped with hand controls and also with a digital recording device. Commercially available lathes are not equipped in this way, so the choice would be between fitting handwheels to an NC lathe, or fitting a conventional lathe with digital recorders. The first option is technically difficult, and both would be expensive compared with facilities provided by means of a programmable controller. They would also pose greater marketing difficulties.

Cutting Technology

The path actually followed in the project was to start from the already-existing programmable controller,[8] and to modify it in ways which seemed likely to meet the 'human-centred' objectives. For ease of programming, the development was carried out in Pascal on graphic terminals connected to a large, central Prime computer. The aim was to do the early, more speculative developments on the Prime, where frequent changes in the software were relatively easy. At a later stage, the software would be rewritten to run on a micro-processor in a programmable controller connected to a lathe, and further developed in this more realistic situation. In fact, the stage of transferring to a micro-processor was not reached during the project, and is being done as part of an ESPRIT project which continues and extends its aims (see Chapter 11).

In the existing controller, software was provided which automatically calculated the best way to remove metal. Here 'best' is to be regarded as a technical term: what is obtained is an approximation to those conditions which give the minimum cost, according to some rather crude formula for defining costs. This 'best' solution has to be found subject to a large number of constraints. The power of the machine must not be exceeded, the cutting forces must not exceed those holding the work in the lathe, nor must they cause excessive deflection of the work or the tool, etc.

The result of these calculations might be that three successive passes of the tool were required (three roughing cuts) each 3 mm deep. The advance of the tool for each revolution (feed) would also be obtained, and the surface speed of the work: say these were 0.3 mm and 130 m per minute, respectively. When the operator was ready to carry out the operation, these values would be used by the machine. In the controller as it existed, they were not communicated to the operator beforehand, and might come as a shock when the machine was started, as being undesirable or even dangerous.

If the operator found that the computed values were unsatisfactory, his scope for changing them was limited. The cutting conditions could be made less severe (smaller cuts, lower feeds) by reducing a 'clamping factor'. This was a number, between 0 and 1, which had been entered by the operator in response to a question when he was defining the area of metal to be removed in the operation: its primary purpose was to deal with situations where the work was not held as firmly as usual. The speed of cutting could also be changed by means of an override knob which is customarily provided on the control panel. If this is set to 100%, the machine uses the calculated values, while when it is set lower or higher, the calculated speeds are adjusted accordingly.

If the operator is to be in charge, rather than subordinate to the machine, this situation obviously has to be changed. The operator must know, before cutting begins, what speeds and feeds and cuts are to be used. He must be able to change the computed values before cutting starts, and must be able to stop cutting at any time, change the values, and re-start. When he changes the computed values he must not lose computer support, because otherwise he will be strongly constrained to follow the path defined by the computer.

He must also be able to obtain easily all the information he needs about the computed results to guide his own decisions.

The details of the calculation which the computer makes, in order to decide the 'best' cutting conditions, can be based on the best existing technological knowledge. This will change with time, and the computer program can change with it. The operator needs to know the basis on which the computer obtains its results, and it is desirable that this should agree with his own understanding of the cutting process. But many of the finer details need not concern him, any more than the details of the calculation by means of which a pocket calculator obtains square-roots. It is the relation between the computed results and the reality of the workshop situation on which he must concentrate, and details become important when they affect this.

One aspect of the calculation does pose new problems, outside the scope of orthodox cutting technology. The operator needs on occasion to substitute his own values for some of those suggested by the computer. When he does so, the computer has to regard the operator's values as fixed, but to obtain the 'best' values it can for those that remain. So, for example, a cutting speed or feed may sometimes be an outcome of a computer calculation, and sometimes may be an input to this calculation. The algorithm in the computer must be able to deal with all combinations of such changes.

Design

The way in which these requirements were ultimately incorporated in the software is described in Chapter 4. The situation described there was reached as the result of a long process of discussion and experimentation, in which the values and aims of the project came into question as well as the details of the technology. Discussions occurred in weekly meetings of the research team and in more frequent informal meetings, in (approximately) quarterly meetings of the Steering Committee, and with operators from industry as described in Chapter 3.

Under industrial conditions, these and similar discussions about other aspects of the system would have had to be truncated, in the interest of completing the design. But one of the aims of the project was to explore the problems of joint work by technologists and social scientists, and if possible to develop a methodology. Discussions were therefore carried on beyond the point where they would normally be terminated.

The unfocused nature of these discussions was compounded by the learning phase of the design team. Any team which is brought together to carry out a design task will take some time to settle down and work successfully. If, as here, most members of the team had no previous experience of design, one would expect this period to be 12 to 18 months, even if all the members had the same background; and when social considerations are added to the technical ones, the period is likely to be increased. Because of the difficulties in recruiting the Research Assistants, all three were in post together for only 26 months. It is therefore understandable that only towards the end of this period was the team working in a fully effective way.

The 'settling down' process in a design team is largely one of reaching an unspoken consensus which guides the decisions to be taken. Even when it is carried out within a single discipline, such as engineering, design cannot be reduced to an unambiguous, methodical process. The book edited by Wechsler[12] contains this description: 'design does not consist in the realization of form, but in the elimination of "misfit". The designer approaches his task with a set of tacit criteria, which appear only when some specific design is found to be inconsistent with one of them. The norm is known only negatively, when it is infringed.'

Design usually starts with a functional specification, which defines what requirements have to be fulfilled without, in principle, constraining the means by which this will be done: in practice a great many assumptions about the means are usually included. Then the team has to generate the most satisfactory design it can achieve to meet these requirements.

To the designer, the process appears to be one of creation, but looked at from the outside, it can be seen as a tree structure of branching decisions.* At each stage, certain technological and social options are open, and a choice has to be made. This will lead to further options and further choices. Each decision restricts us to a smaller and smaller part of our branching tree, and closes off options which previously were open. Some back-tracking is possible, but only within limited bounds of time and resources.

It is not possible to list all the options at once, and make a single global comparison between them, because their number grows exponentially with the level of branching and rapidly becomes very great. Decisions taken early in the design process therefore have consequences which cannot be completely foreseen. They are guided by the collective perception of 'misfit' in the design team, which rests upon a shared background of knowledge, experience and values.

A way of rephrasing the quotation from Wechsler is that the design process is driven by the designers' discontent with tentative solutions. Given some problem to be overcome during the design, tentative solutions will be generated one after another, and rejected if they do not fit. Once a solution has been found which eliminates misfit, it is accepted, and the process moves forward to the next problem and the next decision point: the possibility of further solutions which also eliminate misfit is not pursued. As soon as the designers have eliminated their discontent, they have destroyed their motivation to generate and explore these further solutions.

The rejection of tentative solutions which do not fit is an explicit process, and reasons can always be given which justify the rejection. Acceptance of a tentative solution which is felt to eliminate misfit is much less explicit. The solution will 'look good', or 'feel good', and though reasons for accepting it can be given, these are less influential than the absence of objections. There is ample scope for implicit values to affect the judgements made.

These characteristics of the design process are an obvious response to the exponential increase in options with the number of successive decision

---

*The two ways of regarding the matter are not in contradiction. There is a sculpture by Michelangelo in which a human figure emerges from the uncut stone. In creating the figure, Michelangelo rejected an infinite number of alternative forms which the stone could have taken, and which remain latent in the unworked rock.

points considered. The decisions are made (more or less) sequentially, one at a time, with incomplete information about their consequences. At each decision point, the generation of new options is terminated (more or less) when one satisfactory solution has been produced.

All of this tends to be unfamiliar to social scientists, or indeed to anyone inexperienced in design. A common reaction when faced with a difficult design decision is 'keep the options open'; but this is equivalent to 'stop designing'. The design process is inescapably a process of closing-off options.* A very few of the great multitude of options can, of course, be kept open, and if they are especially important, and doubtful, we should do so. We then need criteria to decide when a decision is so important that it should not be taken during the design process.

If a decision is postponed in this way, there are a number of costs. Experimental work may be needed to investigate the two options, and this may be technical or may consist in obtaining information from users. In either case, two parallel designs may have to be pursued to some degree of completeness before a decision can be obtained. If two options are retained in the final design – for example two different ways of achieving the same end in a software package – the system becomes more expensive, harder to understand, and possibly less easy to use. These costs have to be weighed against the benefits which are sought.

Learning, but Learning What?

The 'blank table option' mentioned in Chapter 1 arose at first from an inability of the design team to reach an agreed solution at a certain point in the development in the cutting technology software. Many such points of disagreement arose, and the 'blank table option' was chosen for deeper study, and as being typical of others.

As conceived originally, the software presented the operator with computed cutting conditions which (with some infrequent exceptions) satisfied all the constraints. The operator was shown these, and each entry was accompanied by a blank space in the table which allowed (and indeed invited) the operator to replace the calculated value by his own. Further details are given in Chapter 4.

The objection raised, from the social science side, was that in order to override the computer, the operator had to have an understanding and skill in the selection of cutting conditions. This skill can be retained only by practice, and if the computer is right most of the time, the operator's skill will decay and he will come to accept the computed value without question.

It was therefore suggested that if he wished, the operator should be able to call up a blank table and enter a complete set of values of his own. Then, again if he wished, he could call up the initial computed values which he had

---

*The sculptor who wishes to retain all the options available in his block of marble must leave it untouched.

not yet seen. By comparing the two he could evaluate his own performance and improve his skill in selecting appropriate cutting conditions. This option would not allow anything to be done which could not as easily be done with the original scheme, but would present a different situation to the operator.

A technical objection is that the existing skill of operators is most reliable in standard conditions. In unusual situations, for example when the work projects far from the chuck, the operator will have to be cautious because he cannot make the calculations which would be required to check deflection, and the holding forces needed at the chuck. Starting with conservative values, the skilled operator, after a few parts, would probably, through successive changes, have arrived at good values. But if the batch is small, a large proportion of it may have been machined with conditions that were far from desirable. An operator who insisted upon experimenting in this way, from a conservative, estimated initial value, would produce less work than if he were prepared, on the basis of successful experience, to accept the computer's initial values.

If the 'blank table' is regarded simply as a teaching aid, however, this difficulty can be avoided. The operator can experiment with the system when his own tasks permit, and while the machine is occupied with an existing job: metal cutting to an existing program can proceed in parallel with programming. Before actually completing a program, the operator can check whether his own values lead to grossly slower machining than the computer's, and increase them if experience has given him confidence in the computed results.

This issue was debated at great length, and details will be found in Chapter 3. Lisl Klein, the social scientist on the Steering Committee, spent a full week at UMIST at a later stage in the discussions, and part of the week was devoted to the 'blank table' issue, and the organisational means for dealing with such problems (Chapter 6). Yet a comparison of the present chapter with the two that follow will show that disagreement remained as complete as before. I agreed that the 'blank table option' should be implemented in the software, because later experience with operators should throw further light on the question. But I was unconvinced by my colleagues' arguments.

The disagreement in fact turned out to be a very basic one, about the aims of the project, and it is reflected in a change of terminology which occurred. My original title for the project was 'A flexible manufacturing system in which operators are not subordinate to machines', but as time went on the phrase which was used to describe the system was 'human-centred'. The second description is much wider than the first. It points accurately to the contrast with Taylorism, but incurs some risk of emphasising the social conditions within manufacturing, to the exclusion of the social environment in which manufacturing is carried on, and the constraints which the interests of this environment impose.

My objection to Taylorism was, and is, that it subjects workers to the organisation (which is the concern of managers and social scientists) and to the machines (which are largely the concern of engineers). I did not, and do not, object to 'building skill into the machines', provided that the machines are subordinate assistants in performing the work. Provided, that is, that they cooperate with the skills and abilities of workers, instead of attempting to eliminate them. The machines should allow existing skills to be used, but should also allow these skills to evolve into new skills related to the new technology. In this process I see it as inevitable that some existing skills will become obsolete, but new skills will arise to replace them if the opportunity for this is provided. In other words I should like to see F4 in Fig. 1.1 in place of the Tayloristic F1.

To some of my colleagues, however, I believe that F4 appeared as Tayloristic as F1, because it accepted that existing skills would, and legitimately could, be made obsolete by being built into machines. So, to them, it was not sufficient that machines should be subordinate to those who worked with them. Machines, they believed, should not be allowed to render obsolete the existing skills of workers, acquired over many years, respected and rewarded, and a source of satisfaction in use. Their preferred view of the future was therefore more like F3, whereas I did not believe that any system designed with F3 in mind would be able to compete with the conventional, Tayloristic development.

Accordingly, they believed that it was important for the operator to be as good as the computer at what the computer did, and wished to see a learning situation for the operator which would achieve this. I was content to see the computer take over from the operator his skill in that routine area with which the computer could cope effectively. Recognising that the computed results would on occasion be wrong, I believed that the operator's skill should in time come to lie in the area where he disagreed with the computer. The learning which I wished to see was therefore in critically comparing the computer's output with the real world of experience, which was available to the operator but not to the computer. An attempt to emulate the computer, I felt, could weaken the informed scepticism about its results which was the operator's strongest potential contribution.

A disagreement at this level is very difficult to deal with. It prevents the consensus which allows a design team to recognise a good solution to a particular difficulty. It ensures that while agreement can often be reached upon general statements, such as the design criteria in Chapter 3, disagreement will arise again about their application to a particular situation. It also produces a tendency for every discussion of particular decisions, which have to be taken in the design process, to wander away into 'generalities' which reflect the basic underlying disagreement.

Discussion of the points of disagreement by the members of the design team among themselves does not seem likely to resolve them. Perhaps we shall have to wait, for a full resolution of the difficulty, until a body of experience in joint social and technical design (and experience by its users) leads to a consensus among its practitioners. It is the existence of such a body of experience, and the common understanding that it produces, which explains the relative ease with which a purely technological design team is able to work.

My Own View

What precedes is an account of some major questions raised during the research, which will not be answered fully without much further study and experience. The concentration on problems should not be taken to imply that progress was not made: towards the end of the project, the design team was working effectively, if more slowly and with more difficulty than would be expected in a purely technical problem. Results from the project have been taken over by the subsequent ESPRIT project (Chapter 11) and rapidly integrated into it.

The account has drawn out some of the differences which arose between my own views and those of colleagues, but no connected account of my own position has been given. It will be briefly sketched here to the extent that is necessary for understanding the differences, and in particular for understanding why I did not accept the social science view as fully authoritative on some of the questions raised.

First, any attempt to redirect technology must satisfy some severe constraints. The rate of progress along the alternative path must be comparable with that along the orthodox path, otherwise anything done will be rapidly outdated. This means that progress of the alternative design must not be too much impeded by the extra social dimension which is introduced. Reasonably fast and effective design procedures need to be developed as soon as possible.

Secondly, the alternative designs must be economically comparable with conventional ones if they are to have any chance of being accepted. Guided by More's *Utopia*, and by *Erewhon* and *News from Nowhere*, we can all imagine worlds in which this constraint does not apply, but here and now it cannot be escaped. It is not too daunting, because conventional technology makes very good use of machines, but very poor use of people. An alternative which makes better use of the resource which is represented by human skill and ability and initiative starts with an immediate advantage.

Thirdly, however damaging are some of the aspects of our modern technology, simple rejection of it is hardly open to us. That would entail a return to the situation where four-fifths of the population lived in poverty on the land, and a further large proportion lived only slightly better as servants or craftsmen. Only a small élite would be able to enjoy leisure and those activities – sports, music, literature, gardening and a hundred others – which a large part of the population now value and pursue.

Because of these constraints, my belief is that the redirection of technology cannot be used to cure the problem of unemployment. This is no doubt the most serious problem which industrialised societies face, but as suggested in Chapter 1 the solution has to be sought in other ways. We have to accept that the necessary work to produce the goods and services required for a life of modest comfort is continually declining. We have to find solutions to the problems this poses, and particularly the problems of inequality, and cannot remedy the situation by artificially increasing again the work required to meet our ordinary needs.

The UMIST project was therefore conceived as a way of attacking one part of a much larger problem: that part which leads to degradation of men and women by treating them as machines in the work that they do. It accepted the constraints of our particular society at this particular time, as being necessary conditions for success, and because success would have a strong positive influence upon the development of technology.

This attitude was certainly criticised in the subsequent ESPRIT project as being too limited, and it was suggested that we should aim at a much more fundamental change, involving not only technology but society at large. Here, I think, one meets a basic division of attitudes. Some will pursue a millennial vision regardless of any calculation of success – and may indeed succeed but more often will fail. Others will settle for what good can with some probability be achieved.

It is good that both attitudes should persist: both are needed. But I accept the second rather than the first. I would prefer a successful project, which changed technology for the better in some definite but limited way, to a gallant failure which aimed much higher. Gallant failures may be admired, but they are not widely copied. Nor do they appeal to funding bodies which are the guardians of pubic expenditure.

Accordingly I urged that we should not indefinitely prolong the design process, should not too greatly complicate the system by incorporating many alternative ways of achieving the same end, and should not attempt to preserve work for people when it could be done completely and just as well by machines. These were the bases of some of our disagreements.

On the other hand, if I saw the aims of the project as circumscribed by the need for success, they go beyond a simple opportunism. The degradation of work may be only one of the objectionable aspects of our existing technology, but it is a highly damaging one. 'It once came into my mind,' says Dostoevsky,[13] 'that if it were desired to reduce a man to nothing – to punish him atrociously, to crush him in such a manner that the most hardened murderer would tremble before such a punishment, and take fright before-hand – it would be necessary to give to his work a character of complete uselessness, even to absurdity.' One has to admit that, without malicious intent, some parts of modern industry have come close to this aim.

In treating people as machines, Taylor proposed to eliminate purpose from their work. They were to work for money, but in work were to be aimless, repeating elementary actions again and again in the way they had been shown. 'The man who puts in a bolt,' said Ford,[14] 'does not put on the nut; the man who puts on the nut does not tighten it.' 'The Ford Company', confirmed two of his followers,[15] 'has no use for experience, in the working ranks anyway. It desires and prefers machine-tool operators who have nothing to unlearn, who have no theories of correct surface speeds for metal finishing, and will simply do what they are told to do, over and over again, from bell-time to bell-time.'

Now an immense amount of knowledge, and skill and purposive action was needed to run the production system which Ford created, far greater than the sum of all that his workers would have needed if they had been employed a century earlier as wheelwrights, carpenters, blacksmiths and turners. The problem with Ford's technology was not that it made no demands on those human abilities which have traditionally been exercised in work. The problem was that the exercise of these abilities had been largely withdrawn from workers and given to a specialised group. In Fig. 1.1, the corona, as well as the sun, was larger than it had been a century earlier, but it was reserved to an élite.

This, as has been argued before, leads to an inflexible production system. Because a great many of the people involved in it are not expected to express any purpose in their work, and are not provided with the means to do so, the response to any unusual situation will be slow and probably inadequate. The passivity which the working situation engenders is damaging to the worker, and destroys the incentive to understand, to develop skill, and to control the work. The project was conceived with the aim of changing this situation.

Finally, we need not be discouraged by the existence of differences of view between those concerned in the project. They are to be expected when people have different backgrounds and experience, and different views on what is desirable in the future. Such differences need to be clarified, discussed, and lived with and worked through. One has to be aware of the danger of 'splitting', as described in Chapter 6, but to attribute all differences to this cause would be to trivialise the discussion, and make it unproductive, by suggesting that there are no substantive questions at issue.

Terminology

In writing, I have hesitated over the name to be given to those who work directly with machines. 'Turner', where it is appropriate, seems unobjectionable, whereas 'worker', 'operator' and 'machinist' all carry overtones of existing, Taylorised working situations. We lack a commonly accepted word for a worker who is expected to exercise initiative and control in the working situation. 'Manager' of an FMS is possible, but has other overtones. 'Controller' of a machine would be precise, but is unfamiliar. The right word will no doubt arise with the right working situation.

In a similar way I have hesitated over the multitude of ways which have been suggested for avoiding the masculine bias of common English usage. Unfortunately all those I know interfere with the concise and accurate presentation of a difficult subject, and I have accordingly followed the longstanding practice.

References

1. L.T.C. Rolt, Tools for the job, 1965 (B.T. Batsford).
2. Fyodor Dostoevsky, The house of the dead, 1861; Everyman edition, 1944, p. 115 (J.M. Dent).
3. Ref. 1, p. 64.
4. Ref. 1, pp. 83–121.
5. James Nasmyth, in Robertson Buchanan (editor), Practical essays on mill work and other machinery, 1841, p. 396 (John Weale).
6. Michael Polanyi, The tacit dimension, 1967, p. 13 (Routledge and Kegan Paul).
7. David F. Noble, Forces of production, 1986, p. 106 onward (Oxford University Press).
8. J. Boon, L. Satine, S. Hinduja and G. Vale, Back to operator control?, Numerical Engineering, April 1980, vol. 1, pp. 27–29.
9. Ref. 7, p. 144 onward.
10. F.W. Taylor, The principles of scientific management, 1911, pp. 35–36, in Scientific management, 1947 (Harper and Row).
11. D. Gossard, Analogic part programming with interactive graphics, Annals of the CIRP, 1978, vol. 27, pp. 475–478.
12. Judith Wechsler (editor), On aesthetics in science, 1978, p. 148 (MIT Press).
13. Ref. 2, p. 25.
14. Henry Ford, in collaboration with Samuel Crowther, My life and work, 1923, p. 83 (Heinemann).
15. H.L. Arnold and F.L. Faurote, Ford methods and the Ford shops, 1919, pp. 41–42 (The Engineering Magazine Co.).

Chapter 3

Towards a Design Methodology:
A Psychologist's View

Martin Corbett

Introduction

The process of designing software for a CNC lathe, when undertaken by a design team, can be viewed as a process of negotiation, i.e. a decision-making process in which multiple parties jointly make decisions to resolve conflicting interests. In this view, the outcome of the design process may be compromise (parties concede along an obvious dimension of some middle ground) or integrative agreement (parties' interests are reconciled) or a combination of the two. In this chapter we shall focus on the process, rather than the outcome, of design negotiation, with specific reference to the CNC lathe software.

Unlike a majority of industrial design projects, the UMIST project did not work towards the fulfilling of specific, rigid design specifications. Rather, the specifications remained fluid as the project progressed and as the means to incorporate human and social criteria into technical and economic criteria were sought. Indeed, the negotiation of the exact meaning of 'subordination' in relation to human–machine interaction (explicitly stated in criterion III of the original research application; see Appendix 2, p.177) was to become thematic throughout the design project.

The Design Context

The conventional engineering approach to the design and evaluation of human–machine systems (such as a lathe) is based upon the functional requirements to be realised with respect to the technological state-of-the-art. In this perspective the human takes over those functions that are either too complex or too expensive to automate, his or her functions being reduced to those functions he/she can carry out better than a machine. Given these constraints, the design process involves a process of 'human–machine comparability' whereby the allocation of functions within the system is governed by a comparison of the human and technical suitability of specific functions.

As Jordan[1] pointed out, such a comparison is made, of necessity, in engineering terms so that human functions are described in mathematical terms comparable to the terms used in describing mechanical functions. Such an approach, however, inescapably leads designers to the conclusion that any time one can reduce a human function to a mathematical formula one can generally build a machine that can do it more efficiently.

The outcome of such a design process invariably leads the system operator into a role of plugging the gaps in the thoroughness of the designer's work. 'On the one hand, as a convenient, movable manipulator, he will have a category of trivial, infrequent action for which automation is unfeasible; on the other hand, as an intelligent data processor he will be expected to respond to ill-structured and unforeseen tasks.'[2]

Such a design process, therefore, may be seen[3] as 'technocentric' in the sense that human factors are compromised for the sake of optimising the technical factors. An alternative design process was sought in an effort jointly to optimise human and technical factors. To this end, the process of 'human–machine comparability' was rejected in favour of 'human–machine complementarity'. In this view humans and machines help each other to achieve an effect of which each is separately incapable, a design process requiring the human–machine system to be analysed and designed in such a way that the human is enabled to interact purposefully with the machine.

The initial social science input to the design process was, therefore, to outline design criteria concerning human and social factors to be viewed alongside technical and economic criteria.

A study tour of institutions and manufacturing companies engaged in similar 'anthropocentric' design projects in West Germany (under the auspices of the New Manufacturing Technologies Programme) revealed[4] the impotence of such criteria in influencing overall system design if they are not somehow reflected in the criteria used to judge overall system properties. The ultimate arbiter in a comparison of two purely technical design alternatives is economics, but very few human factors can be suitably quantified to prove of positive economic benefit. Indeed, as Rosenbrock[5] has argued, conventional industrial accountancy practice views human skill as a cost rather than a resource.

The ultimate arbiters in the comparison of human system design choices are the users. But clearly day-to-day design choices concerning the lathe software cannot be made practically in such a way. The Utopia project in Sweden[6] attempted participative design in such a manner through the incorporation of a number of print workers working half-time on the project. UMIST project funding constraints excluded this option and potential user involvement was necessarily more spasmodic.

Day-to-day design decisions were therefore negotiated between the members of the multi-disciplinary design team itself, each successive phase being presented to a sample of industrial turners and production engineers for evaluation. To enable an integration of social science and engineering perspectives, a list of human-factor criteria was developed to help to establish a common ground between the two disciplines – a 'positive contract zone' in negotiation theory terminology.[7]

Design Criteria Related to the Human as a Component of the System

One of the problems associated with technocentric systems is that their operation often requires skills that are unrelated to existing skills, with the resultant problems of poor transfer of training, and skill under-utilisation. The UMIST system aimed to utilise existing skills and allow them to develop. However, skills analysis as traditionally performed is an analysis of performance rather than behaviour, a measure of the human–machine system rather than of the human in isolation from equipment. Turning skills have changed over time – the development of the slide rest, numerical control of servo motors, and the digital computer have directly influenced the skills involved in lathe operation. Motor skills associated with hand wheels have largely disappeared with the development of NC. Interviews at Rolls–Royce with skilled turners[8] revealed a complex picture of skills development, and the problem to be faced by the design team was how to decide which turning skills to foster in the UMIST system. It was here that the concept of 'human–machine complementarity' became invaluable.

A skilled turner utilises an almost infinite repertoire of behaviour from a finite behaviour experience. This is most pertinent where the changing environmental contingencies alter behaviour generated from moment to moment. To an important extent the skilled operator comprehends and controls the machining process through an internal model of that process, and it is this creative discretionary skill that complements the computer – which is relatively fast, precise and reliable but totally uncreative. Computer software may complement this skill when its codes and strategies fit the needs and skills of the operator.

The skilled use and control of a tool involves imposing one's own definition of reality upon its use and function. The results which emerge during the work process are judged from a reality, which is not determined by the tool but by the user. In this way a tool is always subordinate to (under the control of) the human.

Human control of a system can thus be related to decisions and alternatives that the system design either allows or denies, and the investigation of these decisions and alternatives allowed the production of results of sufficient specificity to be used during the design process. Thus it was criteria related to the system as the operator's environment that were to establish a common base for design decisions.

Design Criteria Related to the System as the Operator's Environment

The idea of maximising an operator's freedom of choice in human–machine interaction is not common amongst engineers. An important aim for the lathe software design was to ensure that the interface does not constrain the number of useful operating strategies available. Operators should have the freedom to shift strategies without losing software support. 'Useful strategies' came to refer to those machining tasks in which choice-uncertainty (i.e. system disturbance) is acknowledged to exist and where the tacit knowledge and skill of the operator are used to avert or correct error.

In utilising the concept of complementarity in conjunction with the analysis of choice-structure, a list of operating activities available in the preparation and execution of machining tasks was drawn up and divided into tasks which are routine and tasks which contain choice-uncertainty.

The complementarity of human and computer numerically controlled (CNC) systems relies on the design of the interface between them, on the medium and type of information and operating strategies that the two must exchange. For efficient system functioning, information should be utilised as near as possible to the point where it is generated and, in the case of disturbances, the information is almost exclusively generated at the CNC lathe. Interaction between operator and machine should be maximised for those tasks open to disturbance, and thus operator-activated (where data manipulation may need to be changed to fit actual, rather than normative, demands). Those tasks that are routine and thus software-activated (e.g. tool files, primitive mathematical functions) only require an interactive screen editor.

All tasks and software functions that were designed during the project were negotiated, analysed and evaluated with regard to four criteria:

1. *Compatibility:* operation should not require skills unrelated to existing skills but should allow existing skills to evolve.
2. *Transparency:* one can never fully control a process without understanding it. The operator must be able to 'see' the internal processes of the software in order to facilitate learning.
3. *Accountability:* software architecture should be self-describing.
4. *Minimum shock:* the system should not do anything that the operator finds unexpected in the light of his or her knowledge of the present state of the system.

In addition, all choice-uncertainty tasks and functions involved four further criteria:

1. *Disturbance control:* tasks that contain choice-uncertainty should be under operator control with computer support.
2. *Fallibility:* the operator's tacit knowledge should not be designed out of the system. He or she should never be put in a position of helplessly watching the computer carry out an incorrect operation that was foreseen.
3. *Error reversibility:* software should supply sufficient information feed-forward to inform the operator of the likely consequences of his or her action.
4. *Operating flexibility:* the system should offer operators the freedom to trade-off requirements and resource limits by shifting operating strategies without losing software support.

(For elaboration of these criteria, see Corbett)[9].

At the general level these criteria helped to ensure that the technical system provides the experience out of which tacit knowledge can be built and machining skills can be utilised and developed. At the more specific level, they aim to maximise the degrees of freedom available to the operator within the constraints of hardware and the limits of the desired system output.

It was generally felt, by the design team, that an exhaustive list of specific criteria would unnecessarily restrict the creative aspect of the design process. However, as the project developed and technical questions grew in their complexity, despite the existence of a 'positive contract zone', the research team (notably myself and Professor Rosenbrock) failed to reach integrative agreement on a number of specific design decisions. Of these, the 'cause célèbre' was the 'blank table debate'.

Heuristics in Design Negotiation: The 'Blank Table Debate'

This interchange arose from a failure to agree upon the presentational form of the cutting technology software (see Chapter 4). Initial suggestions stressed that values for speeds, feeds and depths of cut should be calculated by the software and displayed for evaluation and possible editing by the operator. The alternative suggestion was that the operator should be able, if desired, to fill in his or her own values in a blank table without first seeing the computed values, and then, again if desired, call up the latter for comparison. A written debate was entered into and its later stages are reprinted as an Appendix to this chapter.

The debate was widely publicised amongst fellow researchers and designers, and comments invariably focused on the need for an empirical evaluation of the two suggestions in order to reach a decision. Such an undertaking was rejected for two reasons. First, a thorough empirical investigation would be prohibitively time-consuming given the manpower and time available, and, secondly, if other design choices were to prove equally contentious, there would need to be mechanisms (internal to the design team) for dealing with them that did not require third-party arbitration.

At the heuristic level, given the criteria of 'disturbance control' and 'operating flexibility', the disagreement was non-rational. Yet, in confronting the issue explicitly, the 'positive contract zone' all but disappeared. Negotiation research (e.g. see Bazerman and Lewicki[10]) has examined a number of ways in which systematic biases bound a decision-maker's rationality (e.g. lack of perspective taking, overconfidence in judgement, escalation), but in the case of the 'blank table debate', a period of soul-searching revealed the underlying interplay of that most insidious of biases, the tacit dimension.

Even without a social science contribution, engineering design (and, possibly, design in general) cannot be reduced to an unambiguous methodical process. Harry Collins' research[11] on TEA (transversely excited atmospheric pressure CO_2) laser design reveals the influence of informal, tacit elements within design networks, and his conclusion that 'it is possible to talk about that which cannot be spoken' (p.184) was borne out by our experience.

With characteristic insight, Professor Rosenbrock[12] acknowledges the strong tacit preconceptions that an engineer such as himself may bring into design decision-making. Until the 'blank table debate', neither of us was fully aware of the other's biases. Indeed, it was the negotiation itself which bore out the view that it is only when some specific design is found to be inconsistent with a designer's set of tacit criteria that the tacit dimension becomes visible.[12] Certainly the previously cited criteria aided in moulding the tacit assumptions of the design team, but the negotiation of their meaning in specific design decision-making situations would seem to be difficult to pre-empt.

Towards an Interactive Design Process

Working within a design team involves a learning process, shared by the designers, in which reciprocal perspective taking is developed. If design entails the elimination of 'misfit' in which tacit biases are known only negatively, when they are infringed (Wechsler,[13] p.148) then integrative agreements may well be optimised through a process of 'constructive divergence'.

In the early stages of design, as designers familiarise themselves with their subject and throw ideas around in a fairly unstructured manner, 'misfit' is usually eliminated very quickly within individual minds and a tentative solution is proffered. Such was the case with the cutting technology. The initial divergence process received little analysis and refinement compared with that given to the tentative solution. Indeed, once a solution appears, hindsight may view the initial divergency with a rationality not fully reflected by the actual process. All attention is quite naturally given to the solution.

To expect a designer to explicate his or her tacit biases in a methodical way as this divergency–convergency process unfolds is to impose a structure upon it which may unavoidably interact with that process, thereby changing it. If such a structure is value-free and unbiased, then perhaps there is mileage in such a method. It should be noted, however, that F.W. Taylor believed he had such a structure to overcome the 'chaotic rules-of-thumb' employed by metal workers.

A design process involving 'constructive divergence' aims at the exposure of 'misfit' (before its elimination) through the generation of a number of tentative solutions to all early design choices. Such a process is not pursued as a means to a simple choice amongst those solutions (as reality is rarely so simple as to be reducible to binary choice), rather to expose the constitutive assumptions and epistemological stances (metaphors) upon which the solutions are grounded.

Many of the early design choices in the UMIST project led to tentative solutions which were evaluated vis-à-vis hypothesised conventional 'technocentric' solutions, making choice a relatively simple matter. Based upon our experience of the cutting technology design (and particularly the 'blank table debate'), the design of graphics software and complex profile programming explicitly followed a process of 'constructive divergency' in which a large number of alternative solutions were generated, analysed and evaluated. The process was somewhat laboured and protracted but eventually produced a solution of a highly integrative nature. In this case, unlike the earlier design phase, the 'positive contract zone' itself formed a design specification *based on* negotiation (and the exposure and elimination of 'misfit'), rather than a common *base for* negotiation.

Conclusion

The list of human criteria utilised throughout the project was of undoubted value in aiding cross-fertilisation between the social science and engineering disciplines. They were of more limited value to the generation and evaluation of specific design choices. A process of greater value in the early stages of multi-disciplinary design may be that of 'constructive divergency' – particularly with regard to reciprocal perspective taking and learning. Certainly this process was of great benefit to me, as a social scientist, in unravelling the complexities of engineering design and understanding my own biases. Indeed, the back-to-basics approach of the second phase of the design project enabled me to grasp the technical considerations involved with a clarity which led me to take a somewhat embarrassed backward glance at my early enthusiastic contributions to the technical discussions. In both phases, the patience of my engineering colleagues played an important role in the design process and contributed immeasurably to the exhilaration and enjoyment I felt as a member of the design team.

References

1. N. Jordan, Allocation of functions between man and machines in automated systems, Journal of Applied Psychology, 1963, vol. 47, pp. 161–165.
2. J. Rasmussen, Notes on human system design criteria, IFAC/IFIP conference on socio-technical aspects of computerisation, Budapest, 1979.
3. P. Brodner, Humane work design for man–machine systems: a challenge to engineers and labour scientists, Proceedings of conference on analysis, design and evaluation of man–machine systems, Baden-Baden, 1982, pp. 179–185.
4. J.M. Corbett and R.N. Holden, Report on Visit to W.Germany, Steering Committee paper no. 61, UMIST FMS Project, 1983.
5. H.H. Rosenbrock, Technology policies and options, in The information society: the distribution of benefits and risks associated with microelectronic applications, Proceedings EEC FAST conference, London, 1982.
6. The Utopia Project, Report no. 2, Arbetslivscentrum, Stockholm, 1983.
7. R.E. Walton and R.B. McKersie, A behavioural theory of labour negotiations: An analysis of a social interaction system, 1965 (McGraw-Hill).
8. J.M. Corbett, Interviews report, Steering Committee paper no. 49, UMIST FMS Project, 1983.
9. J.M. Corbett, Prospective work design of a human-centred CNC lathe, Behaviour and Information Technology, 1985, vol. 4, pp. 201–214.

10. M.H. Bazerman and R.J. Lewicki, Negotiating in organisations, 1983 (California, Cape Publications).
11. H.M. Collins, The TEA set: tacit knowledge and scientific networks, Science Studies 1974, vol. 4, pp. 165–186.
12. H.H. Rosenbrock, Engineering design and social science, ESRC/SPRU workshop on new technology in manufacturing industry, Windsor, 1985.
13. J. Wechsler (editor), On aesthetics in science, 1978 (MIT Press).

Postscript

Since the completion of the UMIST project I have continued my research interest and involvement in the design and development of human-centred manufacturing technology and the reader is directed to a number of publications which provide a more detailed analysis of some of the issues raised in this chapter.

For further discussion of the UMIST interdisciplinary design process, see J.M. Corbett, Human work design criteria and the design process: the devil in the detail, in P. Brodner, Skill-based automated manufacturing, 1987 (Pergamon Press); and J.M. Corbett, Strategic options for CIM: Technology-centred versus human-centred design, Computer integrated manufacturing systems, 1988, vol. 1, pp. 75–81.

For details of interdisciplinary design methods employed in other related projects, see J.M. Corbett, Ergonomics in the development of human-centred advanced manufacturing technology, Ergonomics, 1988, vol. 19, pp. 35–39; J.M. Corbett, S.J. Ravden & C.W. Clegg, The development and implementation of human and organisational criteria in CIM environments, in K. Rathmill and P. MacConnaill (editors), Computer integrated manufacturing, 1987 (IFS Publications/Springer-Verlag); and F. Rauner, L. Rasmussen & J.M. Corbett, The Social shaping of technology and work: Human centred CIM systems, AI & Society, 1988, vol. 2, pp. 47–61.

Appendix: Operator Control and Automation

A Foreword

Martin Corbett: A technical system that does not provide the experience out of which operating skills can develop will be vulnerable in those circumstances where human intervention becomes necessary. This viewpoint has become increasingly widespread amongst designers and industrial psychologists in the field of process control automation (not least because of the Three Mile Island experience).

Research on process control operators reveals that an operator will be able to generate successful new operating strategies for unusual situations only if he has an adequate knowledge of the process (e.g. see Bainbridge, 1981). There are two problems with this for 'machine-minding' operators.

1. Efficient retrieval of knowledge from long-term memory depends on frequency of use.

2. This type of knowledge develops only through use and feedback about its effectiveness. There is some concern that the present generation of automated systems, which are monitored by former manual operators, are riding on their skills, which later generations of operators cannot be expected to have. Operators can learn very little about the feel of control from watching an automatic controller at work (Brigham and Laios, 1975), although skills improve over many years of informal on-the-job experience.

> it is important for operators to maintain and develop their feel for process dynamics. As this can only be done by direct involvement in control this suggests that some control functions should be allocated to the human operator in an automated plant even when this is not technically necessary [cf. Jordan, 1963]. One method which has been tried is for the automatic controller (computer) to suggest a control action, the operator assesses it and then presses an 'accept' button. However, when the computer very rarely makes a mistake the operator comes to 'accept' automatically without any check. This method may be good for giving an operator confidence in the computer's control, but it is no help in maintaining his control skill and useless as a way of monitoring automatic control performance. An alternative is for operators to control the process manually for a short period at the beginning of each shift. This is economically worthwhile if the costs of poorer productivity during this period are less than the costs of lost productivity because the operator is not able adequately to take over control or assess failure. The only other alternative is to use special training sessions on high-fidelity simulators, as in aircraft pilot training. (Bainbridge, 1978, p. 260).

I feel that this has direct relevance to the computer control of cutting calculations on our lathe.

Manual control of a centre lathe involves the operator using his knowledge (acquired through *practical* experience) to determine how a workpiece should be set up, which tools should be used, what cutting sequences are best employed, and what speeds, feeds and cuts should be used. On our lathe all this knowledge is still required (whenever a new workpiece is to be turned) except for the latter which will only be required (and therefore used) for approximately 10% of the time it is required on manual and other MDI CNC lathes. Apart from the fragmentation of the operator's mental planning functions that this engenders, the two problems discussed overleaf remain unsolved.

When operator intervention is needed there is likely to be something wrong with the machining process, so that unusual actions will be needed to control it, and one can argue that the operator needs to be more skilled than average. For this skill to develop (especially in future generations of operators) it would seem essential to re-unite all the operator's mental planning functions and allow him to enter his own values for speeds, feeds and cuts periodically. This practice will then place him in a better position to assess computer suggestions against his own practical knowledge. Ephrath (1980) has reported a study in which system performance was worse with computer aiding, because the operator made the decisions anyway, and checking the computer added to his workload. Indeed, in our system it may be better to allow the computer to check the operator's values against physical constraints, thus supplying speedy feedback and eliminating error, when the operator inputs his own data.

Such a modification is easily written into the present software and, aside from the present argument, will give our cutting technology complete flexibility as it allows for fully manual control and fully automatic (computer) control and all possible combinations of these two extremes through our editing facility.

To avoid the addition of further question and answer interactions at the interface the operator could be presented with a second blank table with the option to press 'compute' at any time. If he wishes to enter all his own data he may do so – pressing 'compute' on completion. If he wishes the software to calculate values, he will immediately press 'compute' and the table will be filled. All possible combinations of these two extremes are possible using this facility, thus allowing an operator to pace his own learning and fully exploit his knowledge.

I feel that the present generation of operators, who are trained on manual lathes, may not benefit greatly from this modification. However, if we are concerned with future trends (e.g. CAD linkage to the lathe) then such a modification will be of great benefit to the next generation of operators whose feel for the machining process may be increasingly more indirect.

References

Bainbridge L., The process controller, in Singleton W.T. (editor), The analysis of practical skills, 1978, (MTP).

Bainbridge L., Mathematical equations or processing routines?, in J. Rasmussen and W.B. Rouse (editors), Human detection & diagnosis of system failures, 1981 (Plenum Press).

Brigham F.R. and Laios L., Operator performance in the control of a laboratory process plant, Ergonomics, 1975, vol. 18, p. 53.

Ephrath A.R., Verbal presentation. NATO symposium on human detection & diagnosis of system failures, Roskilde, Denmark. 1980 (reported by L. Bainbridge at Baden-Baden).

Jordan N., Allocation of functions between man and machines in automated systems, Journal of Applied Psychology, 1963, vol. 47, pp. 161–165.

Comment

Howard Rosenbrock: I still cannot generate much enthusiasm for the 'blank table'. I will set out my reservations under a number of heads.

1. I doubt whether your analogy to the process industries, or to aircraft, is appropriate. An aircraft, for example, has no 'stop' button. If a fault occurs in the autopilot, the aircraft must immediately be transferred to manual control, and the pilot must be ready at all times to accept this responsibility. He must also react appropriately to a range of emergency conditions – engine failure on take-off, depressurisation at high altitude, etc. For these reasons he needs to retain and practise his skills continually, whether in flight or by simulation. Similar remarks apply with more or less force to chemical plants, oil refineries, power stations, etc.

A different analogy, more appropriate to my mind, would be to accounting machines. At one time, the entry of items in a ledger, and their totalling, were separate operations. Clerks developed a very great facility in addition, so that they could add up a column of figures as fast as they could read it. When accounting machines came into use, entry of a new item automatically updated the total.

Now it would be quite possible to provide an option on the accounting machine which said, in effect, 'do you wish to see the total, or do you wish to produce it yourself, with the option of comparing it with the machine's result afterwards?' This would be a way of encouraging the clerk to retain his skill in addition, but I think it would not have much appeal. It would be like retaining a skill in using the slide rule, after pocket calculators became a cheaper and better alternative.

2. The difference between the two cases lies in the need for the old skill. If it is needed, it should be retained and appropriate means should be provided for its retention. If it is not needed, then it is pointless to attempt to retain it. We do not need to retain the same skill in point duty for policemen now that we have traffic lights, nor the same skill in mental arithmetic now that we have pocket calculators. It is, of course, good to retain the *ability* to do these things: it is only the level of skill that is necessary which I am querying. At university, I spent many long afternoons in the drawing office gaining competence in designing steam-locomotive valve gear. A colleague took an engineering apprenticeship and spent many months acquiring high skill with a file. Neither of us has used these skills, and both of us feel that the time would have been better spent practising the skills that we do use.

One can argue that the job of the clerk was degraded by the introduction of accounting machines, but this is not an argument for the retention of old skills when they are redundant. Rather it is an argument for a development of the clerk's job which ensured that new skills were provided to take the place of the old.

3. The question then turns on the need for the lathe operator to suggest initial values for cutting conditions. It is fully admitted that initial values (whether obtained from the operator or the computer) may not be satisfactory, and one aspect of the operator's skill is his ability to change the initial values incrementally. This skill needs to be maintained and practised.

In simple situations (short, rigid job, firmly clamped, normal material and tool, cut limited by tool rather than machine power, etc.) the operator's estimate and the computer's will probably agree closely, and both will be a good approximation to what is needed. This, however, is not a very interesting situation, because the conditions will be close to those tabulated by tool manufacturers, and require only memory rather than judgement.

If the situation is less simple, say work which is relatively long and slender, then I believe that the operator will usually be unable to make a first estimate which is close to the allowable maximum. He will probably have to proceed tentatively, starting with relatively light cuts and gradually increasing them until he judges that he has reached the best values. Often he may not carry out this experimentation, because for a small batch-size the gain will not repay the time spent experimenting. If programming is done in the planning office, there is even less chance that the best conditions will be reached. Either way, there is an economic loss.

If our cutting technology program is successful, it should produce on most occasions a first estimate which is close enough to the maximum to make further improvement by experimentation unnecessary. Sometimes the operator may be able to make a significant improvement, sometimes he may have to reduce the severity of cutting, and sometimes he may have to correct for unusual conditions. All of these actions will use his existing skill and allow it to be practised. Sometimes the operator may make the correction before the cut is taken, but often I think the need for correction will become apparent when cutting begins. For this reason it will be important to be able to stop a cut and re-start with different conditions.

4. In summary, the cutting technology program is intended to give, in the more complicated situations, a better starting point than the operator is likely to produce. This is possible because the operator cannot make calculations of work deflection, etc., as the computer can. Further development, from the initial conditions, is entirely the responsibility of the operator.

I see the area where the operator's skill will continue to be needed as the area where the computer model disagrees with the real world. With experience, the operator will find that the computer is wrong in some circumstances (for example, too conservative, or too severe), because it knows less about the real situation than the operator does. When these conditions recur, he can take appropriate action. They will only occur in a small proportion of cases, but the operator will be handling much more work than he did before, so the opportunity to acquire and use skill need not decline sharply.

You, it seems to me, see the operator's skill continuing to lie in the same area as the computer's model, so that he should try to be as good as the computer at what the computer does. This looks backward, rather than forward.

5. Do we wish to use the cutting technology program as a teaching aid? I think we have to assume that operators will have reached an appropriate level of skill before they can accept responsibility for such a system, so we are not thinking of beginners. It would still be possible to envisage the skilled operator using the cutting technology package to improve his own skill in estimating cutting conditions (with the blank table option) so that he can become competitive with the computer. But as I have just said, it is pointless for the operator to compete with the computer in what the computer does best. He should collaborate with the computer, and the computer should collaborate with him, so that together they can do better than either alone.

It seems to me to be sensible to accept the assistance which the computer can give, and to concentrate on the new areas in which the operator's skill will need to develop. There will be many of these, and I will list only a few.

(a) With more severe cutting conditions it will be more important than before to have the best sequence of operations. For example, heavy cutting at the end of the work should be done before the diameter is reduced near the chuck. The selection of this sequence is best left to the operator.

(b) Systematic defects in the cutting technology program will become evident and the operator will learn to anticipate these. For example, since deflections of the work are calculated by dividing it into only a few sections, highly irregular shapes will produce cutting conditions which are less severe than they might be.

(c) Faster cutting will reduce the time which the machine spends working, other things being equal. So more attention will be needed from the operator to organising the flow of work, optimising the sequence of components, etc.

A Postscript

Martin Corbett: Rather than prolong this 'thrust-and-parry' exchange I would like to focus on the core of the engineering–social science dialogue. As I understand it, your philosophy of work design is summed up in the opening paragraph of section (2) in your reply. To paraphrase: if a current skill is needed, it should be retained and appropriate means should be provided for its retention. If it is not needed, retention is pointless. Such a proposition acquires meaning only when one defines 'need'.

The classical technocentric approach defines this 'need' with respect to the technological state-of-the-art – current skills are retained in those areas of work that have, technically, not yet been solved (i.e. reduced to formal description). The extent to which new skills develop is largely dependent upon the thoroughness of the new design. The 'sun and corona' metaphor aptly describes the historical progress of this trend.

A human-centred approach must add social criteria to this purely technical evaluation of skill need. Our debate over the 'blank table option' stems directly from our failure to unite social and engineering criteria concerning cut calculations. We followed a path on which the question 'Do we need a cutting technology?' preceded the question 'Do we need to retain the operator's skill at calculating cuts?' Our answer to the first was an arbitrary 'yes', our answer to the second was a qualified 'yes' because our technology could not be guaranteed 100% reliable (unlike your accounting machine).

Consider an alternative scenario in which the second question precedes the first.

1. Do we need shop-floor skill for cut calculations?
   Yes. It is an area of machining prone to unforeseen disturbance requiring a discretionary skill unsuited to formal description.
2. Do we need a cutting technology?
   Yes. The computer's ability to remember large amounts of data and to make rapid calculations will enable shop-floor skill to develop and protect the operator from error by checking entered data against constraints. The computer complements the operator's skill.

Having made this general decision, the specific details can be worked out through reference to a list of principles (such as the list in my paper[1]) and with regard to economic criteria.

The essential difference between the two approaches lies in the relationship between the scientific knowledge contained within the cutting technology software and the practical (often tacit) knowledge of the operator. The first approach follows a Taylorist path insofar as scientific knowledge gradually supersedes practical knowledge as the prime instigator of work practices (the expanding sun).

The principle of complementarity discussed in my paper[2] implies that scientific knowledge and practical knowledge can peacefully coexist and feed off each other (q.v. Polanyi). The first approach only allows skill (i.e. practical knowledge) to develop when the operator disagrees with the computer (i.e. when scientific knowledge alone is inadequate), implying that the two are mutually exclusive. You imply this when you describe the computer and the operator 'competing' against each other.

Finally, to return to our definition of 'need', the technocentric approach implies a philosophy that views a universal 'need' for the application of science steadily replacing any 'need' for practical knowledge. An alternative path may be taken which attempts to optimise practical knowledge with the help of scientific knowledge, instead of the other way round. Given your writings on the philosophy of science this may be a rewarding path to explore more fully.

References

1. J.M. Corbett, Prospective work design of a human-centred CNC lathe, Behaviour and Information Technology, 1985, vol. 4, pp. 201–214.
2. Steering Committee document, no. 131.

Final Note

Howard Rosenbrock: We differ more than I thought, but we have probably made the differences clear. They seem to me to be philosophical differences, not scientific (i.e. empirically based) ones.

We have agreed at a recent meeting to simulate the 'blank table option', which can be done fairly simply. Having done so, I believe we should accept whatever empirical result we obtain from operators – that is, whether they want it or not.

I am not very happy about the situation for a number of reasons.

1. The decision goes against the weight of existing evidence obtained from our last simulation
2. We have not established any principles which show why the 'blank table option' is more important than a multitude of others
3. Having put the option in, we shall in effect have shifted the burden of proof – can we find reasons for taking it out, rather than can we find reasons for putting it in?

Chapter 4

The Social and Engineering Design of Computer Numerically Controlled Technology

Paul Kidd

Introduction

The traditional approach to the design of engineering systems that attempt to automate a given process is based upon the formulation of technical specifications that can be realised, at a reasonable cost, using current technology. However, owing to the limitations of the technology these specifications very rarely correspond to the totality of actions required to ensure the successful operation of the system. Consequently, a human operator is often expected to 'plug the gaps' in the thoroughness of a design, by performing those operations that cannot, for technical or economic reasons, be automated. He will also be expected to undertake those actions which are required to cope with unforeseen events and system failures (Bainbridge 1983).

The perennial reference to the fact that an operator can override automated functions, or switch from automatic into manual mode, is very often, therefore, a tacit recognition of the incompleteness of a design. Furthermore, the reliance upon human skill is often seen by designers as a defect in the system which further technological developments should allow them to remedy at some time in the future. Operator intervention, therefore, is seen as being necessary but undesirable – a problem to be solved by further technological development.

A number of major problems arise from a design paradigm that leaves humans to 'plug the gaps' in the thoroughness of a design. First, this type of design methodology often fails to place the human in a position where he can adequately cope with the tasks that he has to undertake. The human is expected to deal with unforeseen events and systems failures, but the way the system operates does not provide him with the opportunity to develop and practice the skills which are required to undertake this role effectively (Bainbridge 1983).

Secondly, allocation of functions is based upon the principle of giving to the machine as many tasks as possible, what is left over then defines the work of the human operator. This approach fragments and de-skills work and eventually results in unsatisfactory jobs (Rosenbrock 1982).

Thirdly, the design method incurs hidden costs that become apparent only when the system is operating. For example, the inability of operators to cope adequately with their tasks may result in unnecessary system downtime while awaiting the arrival of expert help that may not be immediately available. Furthermore, systems may not be operating at the optimum level, owing to the presence of disturbances with which the operators do not have the authority to deal, and which again necessitates the use of expertise that may be otherwise engaged or not immediately available (see, for example, Benson 1983). The fact that expert help is required to deal with these situations also implies that indirect labour costs may have to increase (Brodner 1986). In addition, the creation of unsatisfactory working conditions may result in increased labour turnover, absenteeism, etc., and this will lead to further costs being incurred.

Finally, and most importantly, this methodology implies that technological resources are more important to the success of commercial enterprises than human resources. In the longer term, however, the success or failure of any business is dependent upon the degree to which all the available resources are fully utilised (see, for example, Pascale and Athos 1982). Attempting to reduce the role of humans and leaving them only with those tasks which are too difficult or too expensive to automate is, in effect, an under-utilisation of human resources (Rosenbrock 1981). Humans, it appears, are a cost to be minimised, not a resource to be utilised. Such an outlook is economically wrong and can be proved to be so (see Pascale and Athos 1982; Brodner 1984; Precision Toolmaker 1984; Engineers' Digest 1984).

Although many people will see the application of ergonomic, job design and work organisational principles as a way of overcoming some of the problems associated with technology, the UMIST project has been about using the technology, rather than just the interface to the technology and the way that work is organised around the technology, as a further degree of freedom in the design of jobs that are both meaningful and satisfying, and in the creation of an engineering system that does not subordinate man to machine.

Numerical Control Technology

The production of a turned component using a conventional manually operated lathe depends upon the skills of the turner in planning the sequence of machining operations and in determining the machining parameters (depth of cut, feed and speed) for each cut (the planning phase), and upon his ability to respond to machining problems as they arise and also to some degree upon his skill in manipulating the handwheels that drive the cutting tool (the execution phase).

Numerous developments have taken place which have resulted in a separation of the planning and execution phases of the turner's work. Automatic feeds have removed the necessity for continuous manipulation of handwheels, and capstan lathes, which carry a large number of tools that can be brought into play to cut metal in a predetermined way, have allowed planning skills to be transferred away from the turner. Numerically controlled lathes are another manifestation of this trend towards reducing the turner's control over his work.

Numerical control (NC) technology, as it was originally conceived, separated the planning and execution phases, removed from the machinist the responsibility for dealing with machining difficulties, and replaced the handwheels and autofeeds of the hand-operated lathe by closed-loop servo motor drives. Thus, the planning phase was to be dealt with by an office-based programmer whose job would be to determine the sequence of machining operations and the machining parameters required to produce a component. This information would then be stored on a punched paper tape, which would provide the NC controller with the instructions required to drive the servo motors. When complete, this paper tape would be taken for proving and testing on the NC lathe and corrected as necessary in order to ensure the elimination of machining problems. The role of the lathe operator, in theory, is then confined to loading the lathe with a suitable piece of metal, running the prepared machining program, removing the machined component from the lathe, and cleaning away metal cuttings that accumulate on various parts of the lathe.

In practice the operator's role is usually very different. The NC methodology assumes that the machining strategy can be adequately determined by someone away from the machine, and that all machining problems can be identified during the proving and testing phase. The machining process, however, is not well defined and predictable, and hence NC technology relies upon the skills of the operator to cope with unforeseen and unpredictable events in order to ensure successful machining. Many NC controllers, therefore, are provided with override facilities that allow the operator to adjust the programmed machining parameters. The alternative to this is an inflexible and inefficient procedure that requires the machining process to be stopped until the office-based expert can come to the shop floor to sort out the difficulties.

An NC machine tool is, therefore, a good example of a technology designed using a design paradigm that rejects and seeks to replace a machinist's skills, but which nevertheless relies upon that same skill to overcome the deficiencies in the technology. The problem, however, is that by removing the planning work from the machinist, the opportunities available for exercising, developing, and practising machining skills have been diminished significantly. The operator has been placed in a position where, with the machining information in a form that is suitable for a machine to read, the response to unsatisfactory machining conditions is purely retrospective, being based solely upon a reaction to difficult situations after they have actually occurred. There are no opportunities for the machinist to use his skill and experience to foresee potential problems and thus to take action to avoid them.

In recent years computer numerically controlled (CNC) machine tools have started to appear on the shop floor. These machines usually have facilities which allow the programming to be done at the machine rather than in an office, although they can be used in the more conventional way too. Thus, CNC allows the planning and execution phases to be recombined into one job, and this reverses one of the main objectives of the original developments in NC technology, the separation of planning and execution (for a historical account of the development of NC and CNC technologies, see Noble 1986). Implicit in this change in the technology is an acknowledgement that there are large areas of overlap between the planning and execution activities; to achieve successful execution the operator needs control over the planning phase, and satisfactory planning is, in turn, dependent upon experience of the execution phase. In short, to separate these two activities is to create an inefficient and inflexible (and hence uneconomical) method of working.

A shop floor programmable CNC lathe can be seen as an improvement over the conventional NC machine, since it allows a return to the more flexible modes of working associated with the operation of a manually operated lathe. The problem that the UMIST project addressed was concerned with taking this favourable technological change and developing tools that would help to make the machinist's skills more productive.

Some early developments in CNC technology took place at UMIST in the 1970s (see Satine et al. 1980). The system developed at UMIST allowed a machinist to program the machine tool by specifying the machining operations that need to be performed in order to actually produce the component. This particular method corresponds to the way a machinist would set about making the component on a manually operated lathe, and was viewed from the social science position as being a satisfactory method, since it did not create a demand for a totally new skill but rather allowed existing skills to evolve into new skills.

Although this method of programming the machine tool can cope with most of the components met in industry, it does, however, become difficult to use when the component geometry becomes very complex. In these situations an easier approach to programming is to define the component geometry by means of a series of straight lines and circles. Once the component geometry has been defined in this way, the machinist has to specify the machining operations that are required actually to make the component.

The limitations of the original UMIST system can therefore be overcome by providing the machinist with this additional method based upon straight lines and circles. The resulting system would therefore allow the machinist easily to program both simple and complex parts, and allow him to shift to an operating strategy appropriate to the task.

The original UMIST system also incorporated software for calculating optimum cutting conditions. Unfortunately, the machinist is provided with no facilities for checking or editing the calculations, and is therefore in no better position with regard to the machining conditions than the operator of a conventional NC lathe.

This situation was regarded as unacceptable and unsatisfactory, and the UMIST project therefore concentrated initially upon designing a cutting technology system based upon social, technical and economic considerations. It turned out, however, that most of the project was actually spent on this work because of the disagreements between engineering and social science that arose out of the concept of cutting technology software.

The Development of Cutting Technology Software

The overall objective of the work was to produce a system that would be technically and economically as good as any system designed using the traditional design paradigm. In addition, however, the technology was to be designed so as to complement, rather than reject, the machinist's skills and experience. The resulting system therefore had to provide scope for his involvement in the process of calculating machining parameters.

The technical objective of the work was to develop computer software for the calculation of optimum machining parameters satisfying a number of basic constraints on the machining process (i.e. specified limits on workpiece and tool bending; limitations on the available gripping force of the chuck and the available power, torque and revolutions per minute (RPM) range of the machine; specified surface finish requirements of the component; and the need to produce easily disposable metal cuttings (chips)).

The social objectives were less clear at the start of the work, but there was general agreement that facilities for interaction with the software were necessary in order to allow the machinist to take account of the limitations of the mathematical model and data used in the calculations, and to create a situation where he could make use of his machining experience and skill to foresee possible difficulties and take action to avoid them. It was not until the first software prototype had been produced and shown to machinists, that the social criteria became more explicit as the differences between engineering and social science began to emerge.

One obvious way of calculating optimal machining parameters is to calculate the production cost for a large number of machining parameter combinations satisfying the basic constraints, and then to select the combination of parameters that gives minimum production cost. This approach has, however, two disadvantages. First the data required for the calculations are not generally available at the present time. Second, for a machinist, this type of search procedure is not very transparent, in that it is not easy to relate the method to 'rule of thumb' knowledge and machining experience. Consequently, a more transparent calculation procedure was sought.

It is invariably the case that optimum machining parameters are located on the boundary between the region where the basic constraints are violated, and the region where these constraints are satisfied, with the particular location of the optimum point being dependent upon the exact nature of the constraints. Thus a procedure can be devised in which the number of possible machining parameter combinations is reduced by defining a search line along which, starting from the tool manufacturers' recommended maxima, the machining parameters are reduced in small percentage steps until all of the basic constraints are satisfied. The intention of this approach is to produce results that approximate to the optimum values.

Machinist interaction with this cutting technology software can occur at two levels. The first level of interaction is that of the detailed machining parameters calculated by the computer for each cut. This information is presented in the form shown in Table 4.1. Associated with each of the three machining parameters are two columns: the left-hand column contains the numerical value of the particular parameter calculated by the computer for

Table 4.1.a Display for editing calculated cutting parameters

Cut number	Depth of cut		Feed		Speed	
R1	2.03		0.14		202	
R2	2.03		0.14		184	
R3	2.07		0.14		166	
R4	2.07		0.14		148	
R5	2.11		0.14		129	
R6	1.69		0.2		114	
F1	0.5		0.12		110	
	Time	4–23				

Options
1. Change tables   2. Fix value
3. Un-fix value   4. Re-calculate
5. Reset   6. Bar charts
7.   8. Terminate editing

Table 4.1.b Display for editing calculated cutting parameters showing some editing by the machinist

Cut number	Depth of cut		Feed		Speed	
R1	2.03	**2.5**	0.14		202	
R2	2.03	**2.5**	0.14		184	
R3	2.07		0.14		166	
R4	2.07		0.14		148	
R5	2.11		0.14		129	
R6	1.69		0.2		114	
F1	0.5		0.12		110	
	Time	4–23				

Options
1. Change tables   2. Fix value
3. Un-fix value   4. Re-calculate
5. Reset   6. Bar charts
7.   8. Terminate editing

Table 4.1.c Display for editing calculated cutting parameters following editing by the machinist and re-calculation of those values not edited

Cut number	Depth of cut		Feed		Speed	
R1	2.03	**2.50**	0.14	0.1	202	195
R2	2.03	**2.50**	0.14	0.1	184	175
R3	2.07	1.82	0.14	0.16	166	157
R4	2.07	1.80	0.14	0.16	148	139
R5	2.11	1.69	0.14	0.2	129	129
R6	1.69	1.69	0.2	0.2	114	114
F1	0.5	0.5	0.12	0.12	110	110

Time	4–17

Options
1. Change tables
3. Un-fix value
5. Reset
7.

2. Fix value
4. Re-calculate
6. Bar charts
8. Terminate editing

Table 4.2. Display for editing parameters affecting the calculation of cutting parameters

Roughing surface finish (µm)	16.0		Maximum depth of cut (mm)	4.0
Finishing surface finish (µm)	0.6		Maximum feed (mm/rev)	0.6
Clamping factor	1.0		Maximum speed (m/min)	350
Chip ratio, upper limit	20			
Chip ratio, lower limit	2			
Number of roughing cuts	6			
Number of finishing cuts	1		Total machining time	4–23

Options
1. Change tables
3. Un-fix value
5. Reset
7.

2. Fix value
4. Re-calculate
6.
8. Terminate editing

each cut, whilst the right-hand column is where the machinist can input his own choice of cutting parameters, as indicated in Table 4.1b.

The rules for editing Table 4.1 are very simple. First, any cutting parameter specified by the machinist cannot be subsequently overridden by the computer – the machinist's input remains as specified until an instruction is given to the computer (using the function keys on the terminal keyboard) to the contrary. At any stage in the editing process the machinist can instruct the computer to re-calculate the machining parameters, in which case the remaining parameters not yet specified will be calculated by the computer and displayed in the remaining spaces in the right-hand columns, as shown in Table 4.1c (note, numbers in bold represent machinist's editing). The original computer calculations remain displayed in the left-hand columns throughout this editing procedure. Obviously, the machinist can specify as many of the cutting parameters as he so desires.

The second level of interaction occurs via the table shown in Table 4.2. This table shows the recommended maxima for the machining parameters of the tool in use, which correspond to the starting values for the computer calculations. The table also shows the number of roughing and finishing cuts calculated by the computer, the default values of the surface finish specification for roughing and finishing cuts, the upper and lower limits on the chip ratio and the clamping factor. The upper and lower chip ratio limits define a region where good chip breaking characteristics can be expected, and the upper chip ratio limit also defines one part of the search line referred to earlier. The clamping factor provides the machinist with a means of influencing cutting forces (and hence depths of cut and feeds) by effectively reducing, within the computer program, the value of gripping force available at the chuck. Thus the clamping factor allows the machinist to take account of unmodelled effects in the gripping of the workpiece by reducing the clamping factor to some value less than unity (which is the normal value).

The machinist can change a number of parameters in Table 4.2. First, he can edit the default values of the surface finish specifications for roughing and finishing cuts. Second, he can reduce the clamping factor to some value less than unity as described above. Third, he can change the upper and lower chip ratio limits and thus redefine the region where acceptable chip breaking occurs and the search path followed during the computer calculations. Finally, he can also demand a specific number of roughing or finishing cuts. Any changes made by the machinist at Table 4.2 level will cause any previous editing in Table 4.1 to be deleted, and new calculations to appear in the left-hand columns of that table.

The time shown in Table 4.2 corresponds to a cumulative estimation of the machining time for all operations that have been programmed for the given workpiece, whilst the time shown in Table 4.1 corresponds to an estimate of machining time for the current operation.

With the exception of the speed constraint, the computer will generally always produce results that satisfy the basic constraints. A problem with the speed constraint can arise when the radius of the workpiece is very small. In this situation it may not be possible to satisfy the tool manufacturer's recommended minimum cutting speed because as the cutting radius decreases, the RPM of the workpiece has to increase in order to maintain the cutting speed above the recommended minimum. However, when the RPM of the lathe reaches the maximum attainable, it follows that the cutting speed must fall below the recommended minimum if the cutting radius continues to decrease. This situation is illustrated by the results shown in Table 4.1a,b,c. The minimum recommended speed is 170 m/min, but for cuts from R3 to R6 and finishing cut F1, the actual speed is below this limit. The problem can be resolved only by using a more appropriate lathe with a higher RPM limit, or by using a tool with a lower recommended minimum speed.

As soon as the machinist starts to specify machining parameters or the number of cuts, the degrees of freedom available to the computer for satisfying the constraints are reduced. Hence under these circumstances it is possible that the computer may not be able to produce results that satisfy other constraints. When this happened in the first version of the software, verbal error messages were presented to the operator, who then either made changes to the specified values or allowed the calculation to proceed without the particular constraints being satisfied.

During demonstrations of the cutting technology software to lathe operators, a number of important points emerged. First, there was general agreement that the tool manufacturers' recommended maxima for the machining parameters were too conservative, and that an experienced user would sometimes want to exceed these values. Whilst the software had been written to allow the machinist to specify values beyond these maxima, there were no data for cutting forces beyond these values, and consequently the machinist would lose software support under these circumstances. Secondly, the verbal error messages presented to the operator were often confusing, and, if more than one error had occurred, sometimes contradictory. Furthermore, the error messages gave no quantitative information about the amount by which a constraint had been violated. Thirdly, in error-free situations there was no way of establishing how close a given constraining parameter was to its limit.

The first problem can be overcome by extrapolating the tool manufacturers' data, thus giving greater support when editing. However, owing to the need for skill and experience in deciding when to exceed the manufacturers' recommended maxima, the computer is not allowed to exceed the tool manufacturers' recommended values. The second and third problems highlight the difficulties of trying to predict every possible error situation, and also show how difficult it is to make the system transparent. However, these two problems were overcome by the introduction of a display of the constraining parameters, in relation to their respective minima and maxima, in the form of bar charts as shown in Fig 4.1. These bar charts can be called up for any cut and give the operator a qualitative feel for the constraining parameters.

The display of the tables and bar charts was further enhanced by the introduction of colour graphics, which enabled the results of the calculations and the bar chart display to be colour coded. Hence, green is used to indicate that all the constraints are satisfied, yellow is used to indicate that some constraint is violated by an amount up to, say, 25%, orange is used to indicate that some constraint is violated by between, say, 25% and 50%, and red is used to signify that some constraint is violated by more than, say, 50%. Using this approach there was no need to halt the calculation when a

constraint could not be satisfied, since it is evident from the colour of the
displayed results that such is the case. The machinist can then examine the
bar charts to find out which constraints are violated, and then take the
appropriate action.

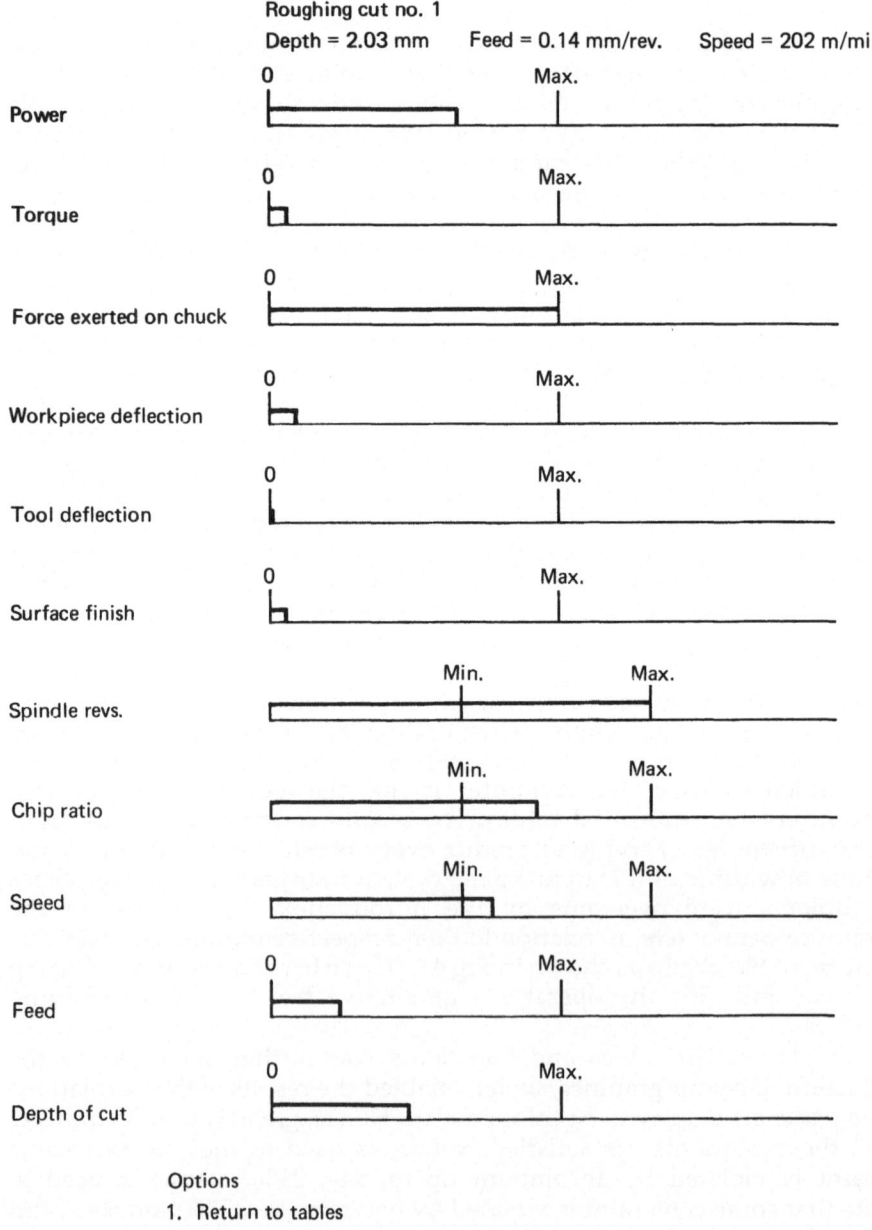

Fig. 4.1. Bar chart display indicating the state of important (constraining) parameters.

The editing facilities thus far described are primarily for use when taking parallel cuts. In some circumstances, however, cuts follow the contour of the desired workpiece shape along the length of the metal (this is known as profiling) and a different type of editing facility may be required in this type of situation. Changing the depth of cut in Table 4.1, changes the depth of cut along the whole length of the cut. When undertaking profiling operations, however, a situation may arise where the machinist wants to change the depth of cut only along a section of the cut. To cope with the profiling situation, graphical editing facilities were developed to allow the operator to edit sections of a cut (see Chapter 5) by redefining the tool path. Using a cursor, the operator specifies a number of points along the required new path, and then the computer draws a smooth curve through these points. Changes in feed and speed can be made via Table 4.1. The operation of the cutting technology software remains unchanged, but the model of the cutting process embedded in the software is more detailed owing to the more complex machining situation arising during profiling.

The 'Blank Table Debate' – A Divergence of Opinion

At the time of writing, it is believed that no commercially available shop floor programmable CNC machine tool system has a cutting technology as sophisticated as the one described above. It is, however, only a matter of time before CNC systems are fitted with cutting technology software. Hence the UMIST project is timely in that it has shown how such technology can be introduced to complement the machinist's skill, and how the kind of problems that arose with the introduction of traditional NC can be avoided by placing the machinist in a good position to cope with disturbances, uncertainty and unmodelled effects. Why then has what is plainly a better approach to design produced such a large amount of disagreement amongst the design team?

Martin Corbett in Chapter 3 mentions that the 'blank table debate' was an interchange that arose from a failure to agree upon the presentational form of the cutting technology. At the heart of the debate was the question of the ascendancy of scientific (explicit) knowledge over practical (often tacit) knowledge and experience.

A common belief in engineering is that technological progress is, in part, about the replacement of 'rule of thumb', experience-based, craft knowledge by explicit scientific knowledge. The development of the cutting technology did not attempt to question or change this belief, but by acknowledging the defects inherent to this scientific outlook, sought to develop a more human-centred approach to design, shaping the details of the technology to the requirements of the machinist.

Thus, the presentational form of a cutting technology that asks the machinist to evaluate and, if necessary, edit machining parameters that have initially been generated by the computer, can be seen as the outcome of a process which sought to improve the existing conventional design paradigm. However, the suggestion that the machinist should be presented with a blank table into which he has to enter some, or all, of the machining parameters, with the computer calculating any parameters not specified by him and checking those values specified against the constraints, is seeking to question the established belief that technological progress can be equated with the substitution of scientific knowledge for practical knowledge. The alternative scenario implicit in the blank table therefore sees scientific knowledge as a means of expanding practical knowledge, and, as such, no conflict can arise between the two, since no attempt is being made by one to reduce the role of the other.

The outcome of the 'blank table debate' was that a blank table option was included in the system. However, both sides of the argument remained unconvinced of the other's point of view and given the different values and assumptions underlying the debate, it is not surprising that this was so. The debate did, however, expose these values and assumptions and also highlighted the difficulties of attempting to introduce social criteria into the design process.

The 'blank table debate' had one further outcome. After many long discussions on the issues surrounding the debate, a more sophisticated version of the blank table was proposed (Kidd and Corbett 1984). The proposal was concerned with the idea of retaining the model of the machining process developed for the cutting technology software, but of using this model in a different way. The alternative approach involves the machinist specifying the machining parameters for one cut, and the model is then used to estimate the constraining parameters, which are then displayed either in the form of bar charts or numerically in tabular form, with the colour coding described being employed in the same manner as before. Six dedicated keys are provided which enable the three machining parameters to be incremented or decremented by a set amount on the depression of the appropriate key. Following any change in the machining parameters, the computer immediately updates the display of constraining parameters, and the machinist can see the consequences that the change in the machining parameter has had upon the constraining parameters. The computer can also advise him on the amount of metal that remains to be removed, so ensuring that the sum of all cuts produced using this method adds up to the amount of metal to be removed.

Now, in the discussions that took place following the formulation of this idea, it was rightly pointed out that using this approach there would be no guarantee that the machining parameters produced would be optimum. However, since there is no conflict between scientific knowledge and practical knowledge in the proposed system, there can be no objections raised to the further refinement and extension of the model of the machining process, since this type of development only serves to improve the quality and accuracy of the predictions made by the computer.

The proposals outlined above were formulated to show that a different kind of technological system can result from a change in the thinking of an engineering designer. However, the proposal did not bring the members of the design team any closer to a resolution of the 'blank table debate'. The proposal served only to highlight further the difficulties arising from individual interpretations of the social criteria listed in Chapter 3.

Conclusions and Personal Reflections – The Engineer as a Social Scientist

When an engineer designs a system to meet some economic objective (an increase in productivity, profit, etc.), he not only designs a technological artefact, but also a social system. Whilst the technical design may in theory be the one which will enable the economic objectives to be met, the social system resulting from the design (and there is invariably such a system) may be such that the economic objectives are not fully met.

The answer to this problem is not for designers to attempt to eliminate the social system or minimise its influence, for that would only exacerbate the problem. Instead, engineers should become social as well as technical designers. This does not imply, however, that engineering or scientific principles should be applied to the solution of social problems – people are not elements of a system nor are they objects to be manipulated like some piece of software or hardware. What is actually meant is that engineers should change their value system, become broader in outlook, recognise the limitations of science and technology, and learn to accept the tremendous ability, intelligence, adaptability and skill that are the hallmarks of humanity.

By developing a cutting technology for a CNC lathe which incorporated social science considerations as part of the design process, the UMIST project has attempted to forge a path that others can follow. The result of the design work has been a system in which the machinist is placed in a good position to make use of new technological developments, whilst at the same time placing in his hands the control, the discretion and the opportunity to exercise skill, which are the prerequisites for dealing effectively with disturbances and uncertainties. In carrying out this work the design team has exposed and come to an understanding of some of the differences between the value systems of engineering and social science, but has not fully resolved the disagreements that arise from these differences.

At a personal level, the day-to-day interaction with the social science colleague on the design team, and the detailed involvement in the design of a technological system involving social considerations, have helped me, as an engineer, to come to an understanding of the problems that the project was addressing. This has also given me valuable insights into the thinking of a social scientist and also into the value system of engineering. Looking back over the project I can now see that my own values have changed considerably, and that this change has been both intellectually stimulating and pleasurable.

My use of the term 'the engineer as a social scientist' is not, therefore, to suggest that engineers should become social scientists. It is more a summary of what, in my own view, are the lessons to be learnt from our work; it is a concise way of saying to engineers that the transition to a different design paradigm involves the realisation of the social nature of their design work, learning to view the technology – and not the human operator – as having limitations, and above all, taking a humane and not a mechanistic view of the operator.

References

Bainbridge, L., Ironies of automation, Automatica, 1983, vol. 19, pp. 775–779.

Benson, R.S., Economic performance control, Transactions of the Institute of Measurement and Control, 1983, vol. 5, pp. 2–6.

Brodner, P., Group technology – a strategy towards higher quality of working life, in T. Martin (editor), Design of work in automated manufacturing systems, 1984, pp. 79–83 (Pergamon Press).

Brodner, P., Skill based manufacturing vs. 'unmanned factory' – which is superior?, International Journal of Industrial Ergonomics, 1986, vol. 1, pp. 145–153.

Engineers' Digest, Flexible but not FMS, Engineers' Digest, 1984, March, pp. 24–25.

Kidd, P.T. and Corbett, J.M., Appendix 3 of Demonstrations of cutting technology software report, Steering Committee paper no. 168. mimeo, 1984, UMIST FMS Project.

Noble, D.F., Forces of production. A social history of industrial automation, 1986 (Oxford University Press).

Pascale, R.T. and Athos, A.G., The art of Japanese management (Penguin Books).

Precision Toolmaker, Precision production uses toolmaking techniques, Precision Toolmaker, 1984, March, pp. 22–23.

Rosenbrock, H.H., Engineers and the work that people do, IEEE Control Systems Magazine, 1981, vol. 1, pp. 4–8.

Rosenbrock, H.H., Robots and people, Measurement and Control, 1982, vol. 15, pp. 105–112.

Satine, L. Hinduja, S. Vale, G. and Boon, J., A process-oriented system for NC lathes, International Journal of Machine Tool Design and Research, 1980, vol. 20, pp. 111–121.

Chapter 5

A Computer Science View

Roger Holden

Introduction

The increasingly widespread use of computers has been seen as heralding a
new 'industrial revolution'. Such a judgement may or may not be valid but
computers will certainly have considerable impact on the nature of work and
our life-style. These changes will no doubt be complex, being advantageous
in some respects, but less so and even positively harmful in others.

The earlier industrial revolution was certainly an event of great complexity
and attempts to sum it up in a single sentence tend to be empty or false. The
great contrasts of the industrial revolution were nowhere clearer than in
Manchester during the 1840s.[1] On the one hand, there was the poverty,
injustice and inhumanity, while on the other hand there was a sense of a new
humanity in the making, a search for justice in the new conditions created by
the rise of industry. Either way Manchester was, to borrow Asa Briggs'
phrase, 'the symbol of a new age'. To some interpreters, Engels and Marx for
example, a new humanity would only arise out of revolution, a throwing
down of the capitalist and a rise of the workers. To others, social, political
and economic reform was needed, exemplified in the Manchester doctrine of
'Free Trade'. The writers of the CSS Report[2] see Mrs Gaskell as not trusting
the 'clear evidence of her eyes' when discussing the social contrasts in *Mary
Barton*[3]. But ultimately, behind the divisions, Mrs Gaskell saw a common
humanity, a more valuable insight than that of Marx who saw little beyond
the divisions.

In judging the 'industrial revolution' it is well to remember how unprecedented were the problems being dealt with at that time. Since around
1740 there had been an accelerating increase in population[4] and it is doubtful
whether traditional industry and agriculture could have coped with this. But
the new industry presented huge problems of management and either
through ignorance or malice, many factory owners sought to solve these by
instituting a reign of terror.[5] Also there is little doubt that much of the
machinery used greatly subordinated the workers to it. Nevertheless, it is

probably wrong to see a single guiding philosophy behind the industrial revolution; utilitarianism was powerful but not all-powerful.

Contrasts of riches and poverty can be found in Britain today, but it would be difficult to argue that they are as violent as in the last century. This perhaps suggests that the image of a new industrial revolution is inappropriate. But in introducing and applying new technology we should seek to maximise the benefits and minimise the negative effects. The particular contribution of the present project is in seeking to reduce the amount of subordination of the worker to the machinery. In pursuing such a project, if we adopt Mrs Gaskell's vision we will look for a wide acceptance, amongst workers and managers, of the system produced. Marx's vision, on the other hand, may lead us to look for a narrow acceptance amongst one particular group.

The project has concentrated on developing software for a CNC lathe and during such a project the computer scientist finds himself developing software to requirements which have been defined by taking into account social criteria, and the software itself has to be assessed according to these criteria. The first part of this paper looks at the structure of software for operator-controlled machine tools, looking particularly at the software developed during the project. The second part then discusses the impact that incorporation of social criteria has on the design process, while the last part looks at the ways in which current developments in computing may be used.

Software Structure

Early numerically controlled, or NC, machine-tools used hard-wired logic controllers, driven by tapes punched with code, known as NC-code, which was interpreted by the controller to produce the requisite motions of the tool. The tapes were produced in an office by specialist part-programmers. With the advent of cheaper computers, the hard-wired controller was replaced by a computer, to give computer numerical control, or CNC. Greater computing power also enabled the development of higher-level part-programming in terms of part geometry or machining operations rather than NC codes, and also enabled programming to be done at the machine. A system combining both of these is described by Satine et al.[6] and is shown to have considerable technical and economic advantages. This can also be seen to be a socially acceptable system, providing a desirable job for the operator, re-integrating the tasks of part-programmer and operator which had been split by NC. For reasons described in Chapter 1, the UMIST project sought to develop this type of system, which could be rendered obsolete by the development of direct numerical control (DNC), where NC programs are produced in a CAD system, or some computer remote from the machine, and sent via a communication link.

A possible software system for an operator-controlled CNC machine tool is shown in Fig. 5.1. The top level is a 'command system' which provides access to the other sub-systems.

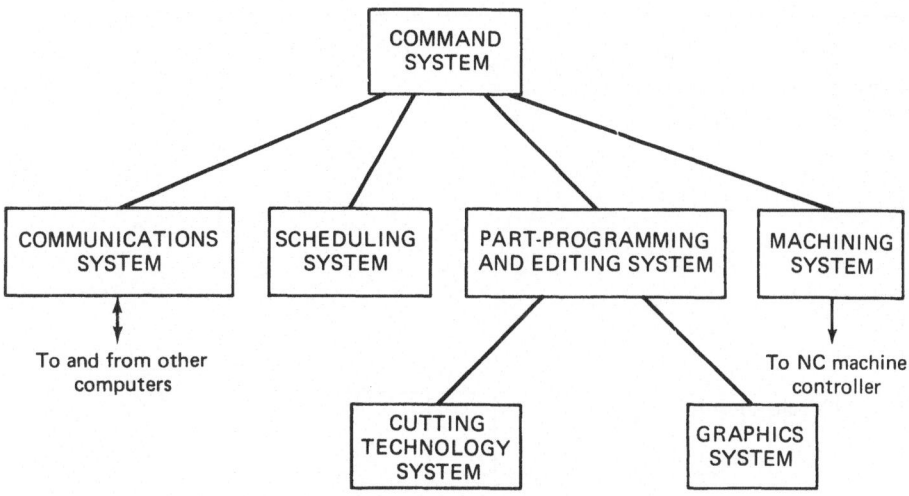

Fig. 5.1. Software structure for an operator-controlled NC-machine tool.

The communication system links the machine's computer to other computers in the factory and could be used for a variety of tasks, such as:

1. Linking to a CAD computer to get drawings of new parts to be machined
2. Link to a computer for storage of part-programs
3. Link to a computer providing scheduling information
4. For returning management information, such as machine utilisation
5. For communicating with other machines, if the machine is part of a flexible manufacturing system.

The scheduling system will use data obtained via the communication system providing for inspection and modification of schedules, if the schedules have been generated elsewhere, or for the actual generation of schedules.

The part-program generator and editor allows for the generation of new, and editing of existing, part-programs. The drawings for new part-programs could be sent from a CAD system, via the communication system, and displayed on a graphics screen, or could be communicated traditionally on paper. Once created, part-programs may be stored locally or transferred to some permanent storage medium or sent via the communication system to a central storage computer. In the latter case, part-programs are then readily available for use by other similar machines in the factory. Part-programs are created and edited using an interactive dialogue, similar to that described by Satine et al.[6] The dialogue system uses terms which are familiar to a machine-tool operator, such as 'turning', rather than abstract geometrical concepts. Part-programs are then stored in high-level form as the answers to the questions during the dialogue. This makes for ease of editing; if only the NC-code is stored, then only this code can be edited, which is more difficult for the user.

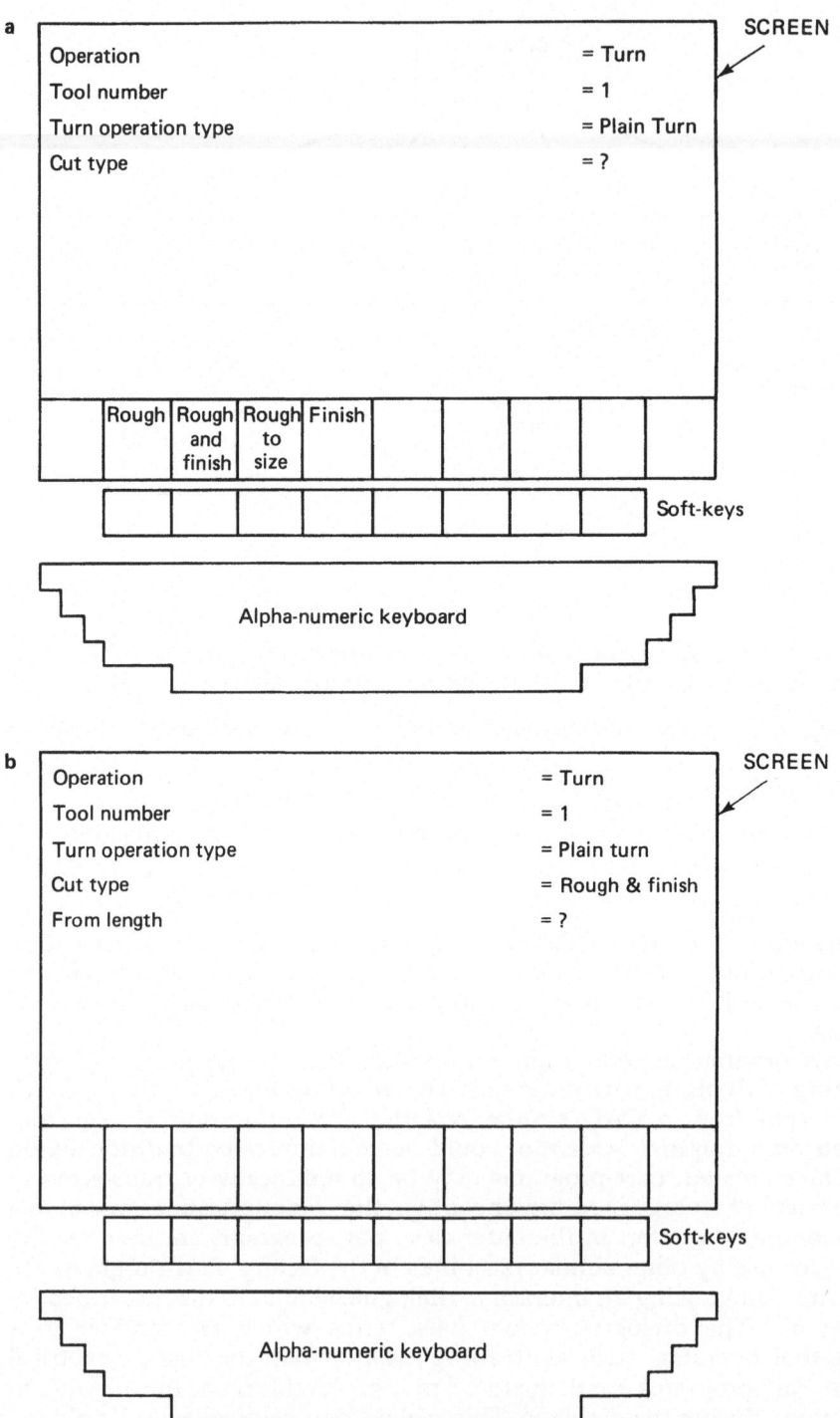

Fig. 5.2. Part-programming. **a** Option selection; **b** numerical input.

The part-program generator and editor contains further sub-systems. The cutting technology is for the calculation and editing of feed, speed and depth of cut. In order to do this, various data relating to the machine itself, the tools and material to be used are needed. These have to be held in a data-base, although tool and material data could be held elsewhere, in a data-base covering all tools and materials held in the factory, and obtained via the communication system.

The graphics system displays the part as it is programmed in order to check the correctness of the geometry being input. In addition it displays tool paths, after their parameters have been calculated by the cutting technology, and allows them to be edited.

Edits to the cutting technology and tool paths have to be stored along with the high-level version of the part-program. This is in contrast to the system of Satine et al.[6] which is purely interpretive: there editing of cutting technology and tool paths is not possible, so they can be recreated from the high-level version each time the part-program is used.

Finally, there is the machining system. This takes part-programs, created by the part-program generator and editor, and produces the NC-code to drive the actual machine-tool controller. There is no need in this system to keep the NC-code; it is created each time the part-program is run.

Aspects of the UMIST System

Although the original intention of the UMIST project was to develop software along the lines described above to control three machines in a flexible manufacturing cell, constraints on time and resources meant that only certain sections of the software for a lathe were tackled. It is hoped that what has been done would be adaptable, without fundamental alteration, to other items of machine tooling.

As mentioned above, the system of part-programming has been derived from Satine et al.[6], where the operator is presented with a series of questions to which he responds by typing in a number or by selecting an option by pressing the designated key. This is illustrated in Fig. 5.2, where in (a) the operator is being asked to respond to an option, while in (b) he is being asked to input a numerical value. Option selection requires the availability of programmable, or 'soft', keys arranged along the bottom of the screen, as shown in Fig. 5.2, although for development purposes the auxiliary key-pad of a VT100-style keyboard was used.

To edit part-programs, a similar screen editing process is used. One operation at a time is displayed on the screen, and there is a cursor which can be moved up and down from line to line, as shown in Fig. 5.3. The cursor is moved to the line to be edited and the 'change' key pressed. If it is a numerical value being changed, then the current value is erased and a question mark displayed; the operator then types in the new value required. If it is an option being changed then again the value is erased and a question mark displayed; the possible options are shown above the keys and the new one is selected. Then all the questions to the end of the operation are re-

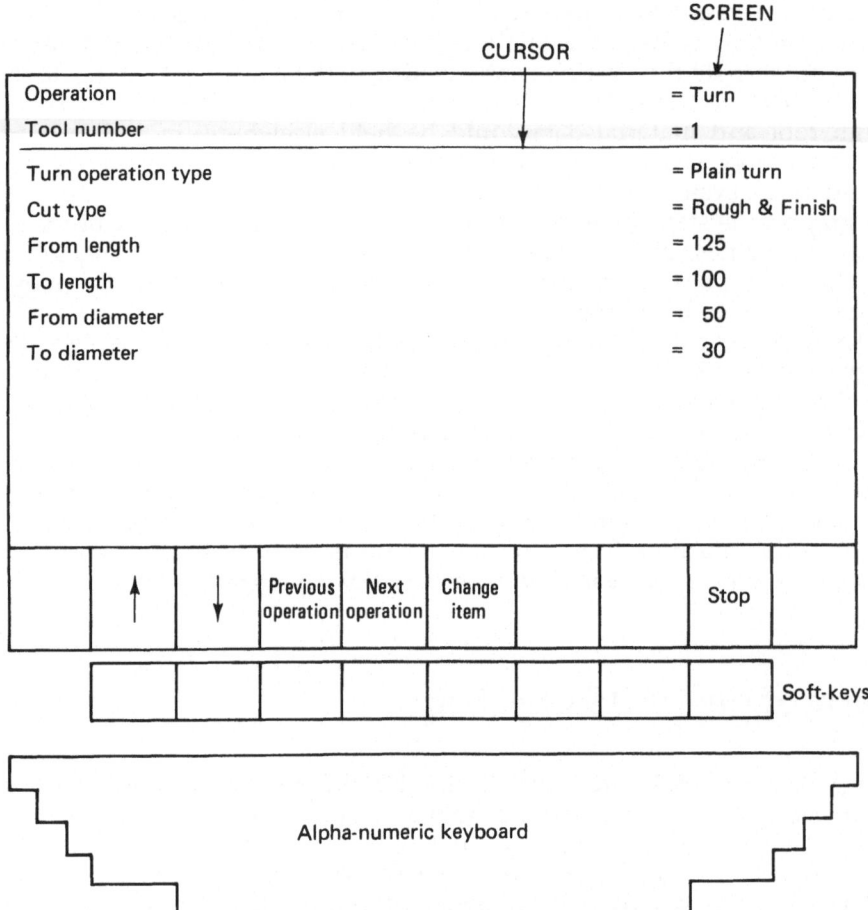

Fig. 5.3.   Part-program editing.

asked, because the option selected defines what questions are to be asked next and the previous ones may be inappropriate.

A skeleton syntax for part-programs is given in the Appendix to this chapter. This defines the questions to be asked and the order in which part-programs are to be stored. From here it will be seen that each part-program consists of a block of data called 'set-up' followed by data for any number of operations. 'Set-up' defines how the workpiece is held and its dimensions. Where a syntax item contains a number of alternatives, for example 'units' or 'work-type', then an option question to select one or other of these alternatives is called for. Otherwise the question requires the input of a number, as, for example, for 'clamping factor'.

Parts of the cutting technology system have been implemented, and indeed this has been one of the major aspects tackled by the project. Little will be said about it here because it is discussed in greater detail in Chapter 4, from which it will be seen that a screen editing approach has again been adopted to allow the operator to modify cutting parameters. A cursor is

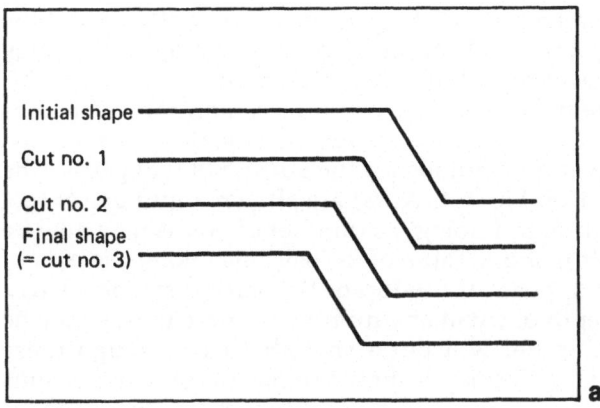

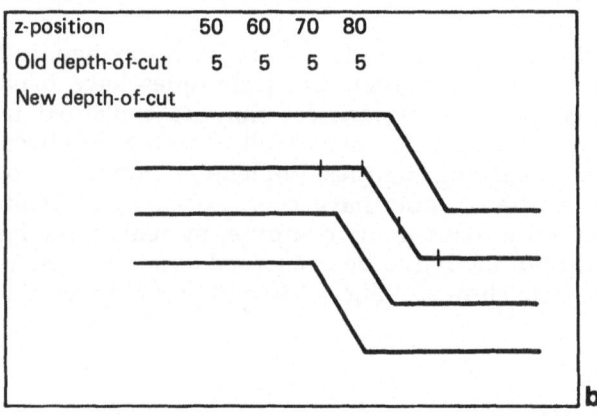

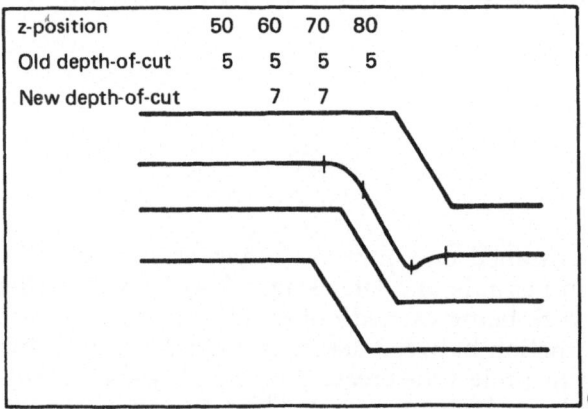

Fig. 5.4. Tool path editing. **a** Initial tool paths; **b** points marked on tool path; **c** new tool path.

moved to a location on the screen representing the value to be altered and the new value typed in. Graphical methods of representation are used to convey information regarding the severity of the cutting conditions.

A complete graphics system for drawing workpieces and tool paths has not been produced because such systems are widely available commercially. Instead attention has been paid to methods for the editing of tool paths. The method evolved is illustrated by Fig. 5.4. At (a) are shown a number of tool paths as produced by the cutting technology; a number of points may now be specified along a cut, as in (b), and a Table of z-positions (along the axis of the lathe) and depths of cut appears at the top of the screen. By means of a cursor, new values for the depth of cut at any number of these points may be typed in, and finally the computer will put a smooth curve through these using spline functions, as in (c). Display technology places some limitations on such a process. For cheapness, a colour raster graphics screen was used for development; this can be very slow at times because of the need to erase parts of the picture before redrawing and the necessity to plot complex curves point by point. The final system, using special purpose hardware, will be much faster.

Despite a potential for the use of declarative ('artificial intelligence') programming techniques, traditional programming techniques have been used in implementing the system; the reasons for this are explained in greater detail below. The language used is Pascal and full advantage has been taken of its data and program structuring facilities. Implementation in one of the older languages, such as Fortran, would have been extremely difficult. Development work was done on a main-frame computer system, allowing free development of ideas without having to be concerned about efficiency. Full implementation of the system shown in Fig. 5.1 would probably require a 32-bit micro-computer.

Software Development

The software development process can be considered to have the following stages:

1. Requirements definition
2. Formal specification
3. Design
4. Program (i.e. code)
5. Test and verify

Ideally, these would be performed one after the other, as suggested by Fig. 5.5a. In practice this is impossible and later stages feed back to earlier ones, as in Fig. 5.5b, some work being carried out on all stages in parallel. This also means that the distinction between stages is not always clear, but generally speaking the larger the project the greater the formality and distinction between these stages, each one being represented by a distinct document.

The requirements definition is produced by the 'customer'. 'Customer' must here be understood in the wide sense of whoever desires the software to be produced: some software is produced to the requirements of a specific

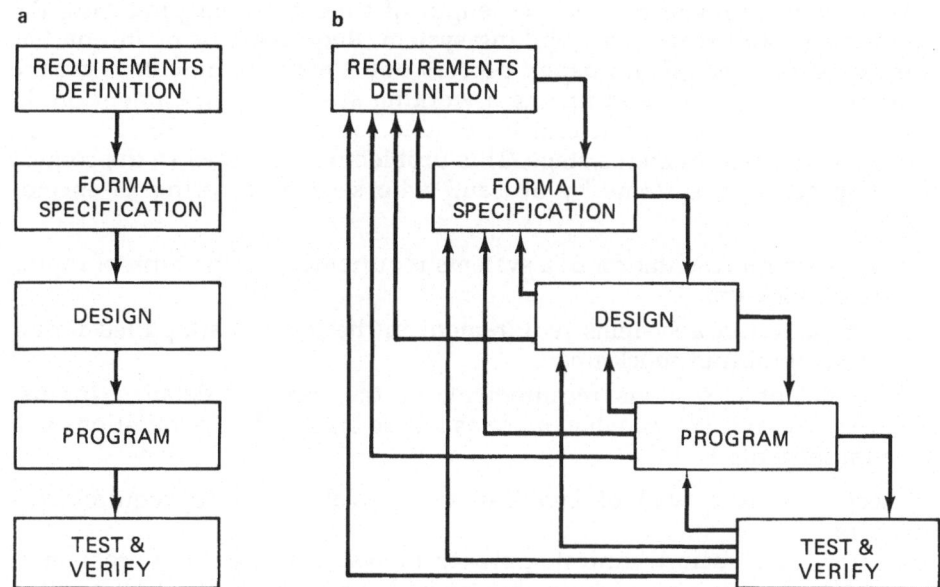

Fig. 5.5. The software development process. **a** Ideal; **b** real.

customer; other items are produced by a company to sell on the open market, in which case the company producing the software develops the requirements itself. Either way, the requirements document will state informally, in natural language and diagrams, the characteristics required of the software system. This may, in fact, be part of the design of a larger system. For example, the requirements for an operator-programmed CNC machine tool will state things like 'a dialogue part-programming system is required'; 'a cutting technology is to be included which allows operator editing', and so on.

The informal requirements document can then be transformed into a formal specification; for example, a formal syntax can be developed for the dialogue part-programming system. Recent work in software engineering has stressed the importance of this stage (see Cohen[7] for example) because it enables ambiguities, inconsistencies and incompletenesses in the requirements to be detected before implementation is reached, when they are more expensive to correct. This process may then require alterations in the requirements.

At the design stage the formal specification is taken and data and module structures for the program developed. This will involve resolving questions such as whether the syntax in our example should be implemented as a recursive descent or a table-driven system.* Then the modules can actually be programmed and the final stage of verifying, testing and debugging embarked on. This final stage may require a return to any of the earlier stages; in particular, when the customer sees the system working, he may realise that the requirements were defective.

---

*These are two methods of writing syntax analysers; for a description, see Wirth[12].

An evident problem here is the length of time between producing the requirements and seeing the working system. Requirements definition has to be done as a 'pencil and paper' process but it may be difficult to get the requirements right without having a working system to experiment on. In other words, it would be useful to be able rapidly to develop prototype systems at the requirements stage. This problem is discussed in the context of data-processing systems by Strelau,[8] who sees prototyping as encompassing:

1. A skeleton representation of a systems requirement in the form of input/ output mock-ups
2. A simulation of a systems requirement in the form of simplified input/ processing/output mock-ups
3. A model of a systems requirement, in the form of detailed inputs/ processing/outputs, capable of use as a base on which live variations will be implemented

He sees this as a way of enabling user involvement in requirements definition.

In pursuing a socio-technical approach to system design, the problem of prototyping becomes more acute, the detailed evaluation of a prototype often being a key element in the process (see Shackel and Klein,[9] for example). In addition, several different requirements may be produced for evaluation, for example cutting technology with or without a blank table (see Chapters 3 and 4). Much socio-technical design has involved existing systems which are to be re-designed or replaced, such as the replacement of a paper system by a computer system. Here there are staff available who understand the system and its objectives, and hence can readily evaluate design requirements and prototypes. With the UMIST lathe project this was not the case; the situation was more like one where a company is developing a system for the open market.

So instead of presenting requirements and prototypes to the users for evaluation, representative workers were used from the types of businesses that were envisaged as using such a system. To do this, fairly complete and highly developed prototypes were needed; for example, to evaluate the cutting technology a full screen editor and cutting technology computation system had to be developed before any operator evaluation could be attempted. Even so, a full system was not presented to operators: it only allowed for plain turning, did not produce NC-code and was not connected to the lathe. Therefore it was shown to operators on a terminal in an office and some operators, particularly those not familiar with CNC, found this difficult to relate to; they would have been happier if the system had been connected to a lathe in a workshop. Unfortunately, the facilities were not available to do this, and additional manpower would have been needed to develop NC code-generation routines.

Two solutions to this problem may be envisaged:

1. The development of software prototyping techniques and tools to enable mock-ups to be evaluated during requirements definition and generally to reduce the time between requirements definition and finished product
2. The development of socio-technical methods so that social problems may be solved with less dependence on the evaluation of prototypes

The UMIST project has tended to concentrate on the second of these in seeking to develop design criteria (see Chapter 3). The first is a part of a general problem in software engineering of which there is a widespread awareness, although in a narrower context than here. Hence future developments in software engineering can be expected which will help to solve this problem. But application of these will reduce the development time of non-socio-technical projects as well, so continued work in the area of (2), above, is important.

Use of Alternative Programming Methods

As noted above, the project has used traditional programming techniques. But adoption of more recent programming techniques may enable systems to be constructed which more closely satisfy the aims of the project. These programming techniques are those commonly referred to as 'artificial intelligence' techniques. They will be referred to here as 'declarative programming', a term which avoids the potentially contentious philosophical overtones of 'artificial intelligence' and emphasises the pragmatic considerations involved in arguing for their use.

There are two types of declarative programming:

1. Functional, or applicative, programming represented by the language Lisp (see Winston and Horn[10])
2. Logic programming, represented by the language Prolog (see Clocksin and Mellish[11])

Lisp is the older of these, but Prolog is becoming more widely accepted and future language developments based on Prolog are to be expected.

Declarative programming techniques are claimed to enable construction of software which:

1. Is compatible with human cognitive processes
2. Can provide explanations

Both of these can be seen as helping to fulfil the aims of the project in producing a system which is subordinate to the operator and satisfying the principles outlined in Chapter 3. The operator would be able to follow the processes of such a system, requesting explanations where necessary, thus enabling him to direct and intervene purposefully in its operation.

However, these techniques were not used because investigation showed that they were insufficiently developed, and that there was a lack of completed work which could be applied to the project. In particular, major parts of the system as described above are numerical, particularly the cutting technology, and today declarative programming techniques are ill-suited to numerical processing. Scheduling is an area where declarative programming may be very applicable, but this was not dealt with during the project. Thus it appeared that use of declarative programming would have led to engagement in research outside the aims of the project.

However, with hindsight it is now possible to outline the type of system which could be constructed using declarative programming methods. The present system, as described above, uses a menu-driven dialogue method with screen editing. At each stage the operator drives the system by replying to a question by either typing-in a character string or selecting one of a number of options. To edit an item in a part-program a cursor has to be moved to a position on the screen representing that item, movement of the cursor being one of the possible menu selections. To do this the program has to maintain a pointer into the data-base, pointing to the same item as does the cursor on the screen. So, for example, if in a part-program called 'Fred' it is desired to change the tool number for operation 3 from 5 to 7, the following sequence is performed:

1. Press key to select 'edit part-program' option
2. Next question will ask for the part-program name, which will be replied to by typing 'Fred'
3. Part-program 'Fred' will be found and its set-up data displayed on the screen
4. To get to operation 3, press 'next operation' menu key three times
5. Operation 3 will now be displayed with the cursor pointing to the first line
6. Move the cursor down until it is pointing to the line labelled 'tool number'
7. Press menu key 'change item'
8. The existing number '5' will be erased and a question mark displayed
9. Type in new number '7'

This takes around 13 key strokes and could take some time to perform, particularly as at step (4) each operation passed through has to be written on the screen.

Now, using declarative programming, it is possible to envisage a linguistically driven system where, instead of following the above sequence, a sentence would be input, thus:

'In operation 3 of "Fred", change the tool number to 7.'

A natural language processor would then parse this sentence to decide what to do, i.e. find part-program 'Fred', go to operation 3, find the tool number and change it to 7. When editing, of course, the operator is likely to want to see the operation on the screen, so, if operation 3 of 'Fred' is already displayed, it would only be necessary to type:

'Change tool number to 7.'

A sophisticated language processor would be needed so, for example, all of the following would accomplish the same thing:

'In "Fred" alter the tool number of operation 3 to 7.'

'Please alter the tool number to 7 in operation 3 of "Fred".'

'In operation 3 of "Fred" change the tool number from 5 to 7.'

The language processor would have to be able to detect ambiguous and incomplete instructions and also check spellings. The spelling checker could assume meanings for simple mis-spellings, but would have to ask the operator for confirmation in more complex cases. Suppose, for example, the part-program name is typed in as 'Free' instead of 'Fred' and suppose that there is no part-program called 'Free' but there is one called 'Freed'. Then the computer could reply with:

'No part-program called "Free", do you mean "Freed"?'

to which the operator could reply with 'No, "Fred"' or simply 'No' in which case he may be asked:

'Do you mean "Fred"?'

and the reply can be 'Yes'.

The process of part-programming may require prompting, although sentences such as:

'Plain turning, from length 150, to length 25, from diameter 100, to diameter 50'

could be typed in. The system would then prompt for additional information such as type of cut, tool number and features at corners.

At present, it is difficult to see how some parts of the system could be implemented in ways other than with conventional, numerical, programming methods. This is particularly true of the cutting technology which has been specified as a numerical algorithm. However, further investigation may lead to ways of re-specifying this in a form more suitable for declarative programming. Indeed, the cutting technology algorithm does include rules as to what is to be done in given circumstances and is one area where explanation of the software's actions is required by the operator. Some arithmetic will always be necessary, but this can be expected to become easier with future developments in declarative programming.

The advantage of such a system would be in terms of flexibility, both in terms of expression and in terms of the range of instructions which could be given to the system at a given time. For example, while inspecting the drawing of a part-program on the screen, the operator may notice a wrong dimension. To correct this in the present system the operator would have to go through a number of actions to obtain the required section of the part-program on the screen, change the dimension and then return to the graphic picture. In the proposed system he could give the editing command without losing the picture. Again, for example, while editing a part-program, instructions could be given to send or receive data through the communication system, or to run another part-program, assuming a suitable processor structure. The only time the operator would be constrained in the nature of the instructions he could give would be when the system was requesting a response to a given question, as during part-programming for example.

Various objections could be raised to such a system. First, it requires more key strokes to perform an action than the present system; in the first example, around 55 key strokes are required as against 13. Second, and perhaps more importantly, it implies a high degree of linguistic ability on the part of the operator, who has to put his desires into a grammatically correct and unambiguous form. The system would prompt for things not understood, but an operator might become frustrated by continual prompting. Also, so far as giving explanations is concerned, it is known that purely verbal explanations are inadequate in describing the actions of the cutting technology and that graphical methods are needed. So further thought and experiment are necessary to devise a more symbolic system that allows for flexibility without requiring sophisticated linguistic abilities.

Finally, to implement such a system may be difficult at present because it would require:

1. An extremely sophisticated natural language processor
2. More advanced declarative languages which are efficient at arithmetic
3. Possibly a processor with a highly parallel architecture

Conclusion

Assessment of this project could take place in two contrasting frameworks, the 'radical' and the 'reformist'. The 'radical' would see a predominance of bad effects in present technology and would therefore look for some completely new direction for technological design which would remove these bad effects. To revert to the imagery used in the introduction, the vision of the radical is closer to that of Marx than that of Mrs Gaskell and looks towards some final solution of the problems.

The 'reformist', on the other hand, sees contemporary technology as a more complex phenomenon, involving much good as well as bad. He is looking for ways of reducing the bad and enhancing the good, recognising that this is a continual process without any final, Utopian state to be reached. The problem will be seen more in terms of lack of knowledge and understanding. Of course there are many shades of opinion between these two.

To the present writer the 'reformist' view is the more meaningful of these two. The project has achieved an understanding of the problems involved in incorporating sociological knowledge into engineering design and has sought a solution to these.

References

1. Asa Briggs, Manchester, symbol of a new age, in Victorian cities, 1968, Ch. 3, pp. 88–138 (Penguin Books).
2. CSS Report, New technology: society, employment & skill, 1981, p. 93 (The Council for Science and Society).
3. Mrs Elizabeth Gaskell, Mary Barton, Ch. 3, 1848.
4. F.A. Wrigley and R.S. Schofield, The population history of England, 1541–1871, 1981 (Edward Arnold Ltd).
5. S. Pollard, The genesis of modern management, 1965, Ch. 5, (Edward Arnold Ltd).
6. L. Satine, S. Hinduja, G. Vale and J. Boon, A process-oriented system for NC lathes, International Journal of Machine Tool Design and Research, 1980, vol. 20, pp. 111–121.
7. B. Cohen, Justification of formal methods for system specification; software & microsystems, 1982, vol. 1(5), pp. 119–127.
8. F. Strelau, System prototyping, in P.J. Wallis (editor), The software development process, State of the Art Report (Pergamon Infotech Ltd), 1985, vol. 13(2), pp. 129–135.
9. B. Shackel and L. Klein, Esso London Airport refuelling control centre redesign – an ergonomics case study, Applied Ergonomics, March 1976, vol. 7.1, pp. 37–43.
10. P.H. Winston and B.K.P. Horn, Lisp, 1981 (Addison-Wesley Pub. Co.).
11. W.F. Clocksin and C.S. Mellish, Programming in prolog, 1981 (Springer-Verlag).
12. N. Wirth, Algorithms + data structures = programs, Ch. 3, 1976 (Prentice-Hall Inc.).

Appendix: Skeleton Syntax for Part-programs

In the syntax diagrams which follow, terminal symbols appear in capital letters thus:

CHUCK

Other symbols are non-terminals, but only a skeleton syntax is given here and not all are expanded. For an explanation of syntax diagrams, see Wirth[12].

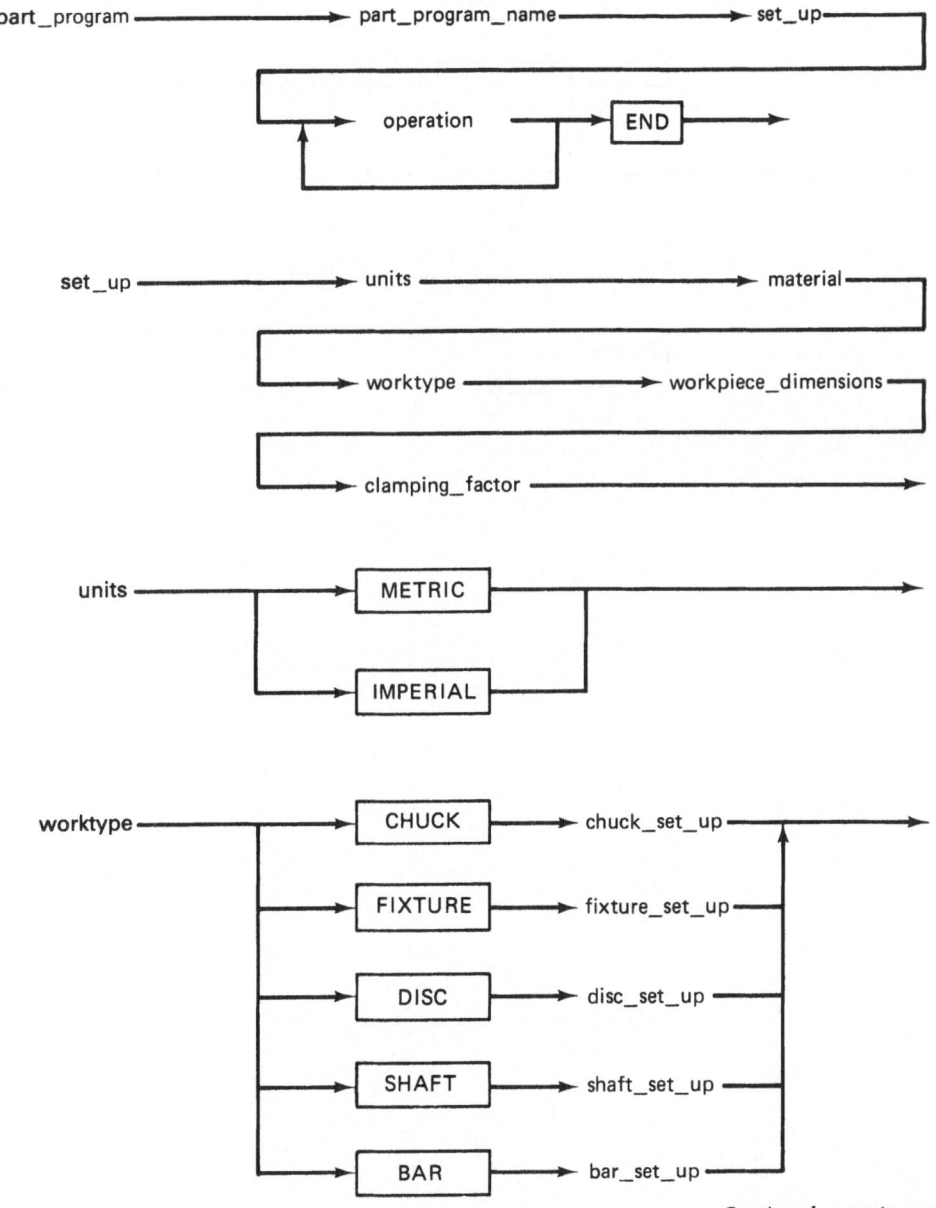

Continued on next page

clamping_factor ──────────────────────── real_number ──────────►

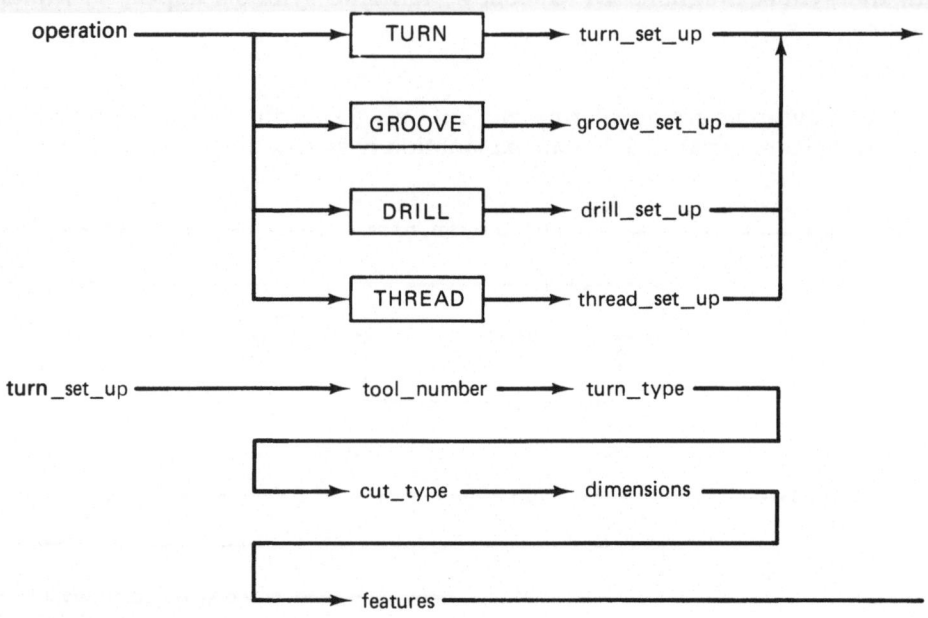

Fig. 5.1A. Syntax diagrams.

Chapter 6

On the Collaboration Between Social Scientists and Engineers

Lisl Klein

Background – The Dynamics

There is a dynamic which psychologists call 'splitting'. It is a process of psychic economy, whereby people tend to simplify a complex situation for themselves by attributing all its x characteristics to one of a pair, and all its y characteristics to the other. The goodies are all-good and wear white hats and the baddies are all-bad and wear black hats, and possibly also black moustaches. Splitting means that one is most unlikely to see a black moustache under a white hat.

Splitting is very pervasive – people identify one football team, one political party, one nation as all-good, and (with a strong tendency to see only two even where there are more) another as all-bad. Scientists are supposed to be all-rational, and artists all-intuitive, industrialists concerned only with money (and nothing else), academics only with knowledge (and nothing else). And, although many people know with a part of their mind that things are really not like that, once splitting is established and becomes institution-alised, those involved get caught up in it and it becomes very hard to break out of. Companies have to encapsulate their 'soft' aspects in personnel departments or donations to ballet companies, in order to maintain their required 'toughness' intact. Politicians are unable to say *anything* good about the policies of their opponents. In turn, people begin to live up to what is apparently expected of them.

Something very similar has happened in relation to technology and its human inputs and outputs. Clearly, they are interdependent: on the one hand, the inputs to design decisions in manufacturing systems are not only knowledge about the properties of materials and the dynamics of machining. They are also, first, factors affecting the individual designer, such as values and assumptions about how people function and about what is really economic and, secondly, organisational factors affecting design processes such as pressures on the team from outside, status differences when alternatives have to be selected, career development issues, etc.

On the other hand, the outputs from design decisions in manufacturing systems include effects on the perceptions, attitudes, skill repertoire and behaviour of individuals, on organisation and therefore also, in aggregate, on society. The consequences for the people who work with and around the technology mean, in turn, that technology is often not operated in the ways in which its designers, from a split position which blanks out the human and social aspects, intended. A split position would then lead one to conclude from this that people should be eliminated from the system, not that they should be taken into account more realistically.

In the context of the UMIST project the human outputs from design decisions in manufacturing systems do not need much documenting. The profound effects of technology design on the work experience, and therefore on the development and autonomy of the individuals and groups who work with it (and thus on society) was the raison d'être of the project in the first place. A shared acceptance of the importance of this held together a very diverse Steering Committee.

What does need pointing out here is that the social and technical aspects of technology are not only split off, but that the splitting against which attempts to work in an integrated way now take place is deeply institutionalised. It permeates education, research, and the professional institutions and their literature. There are populations whose horizons are dominated by the one, and populations whose horizons are dominated by the other. Social scientists read what social scientists have written, engineers read what engineers have written.

One small anecdote may serve to illustrate how this institutionalisation in turn perpetuates the splitting: after some time as a member of the Steering Committee of the project, I considered that I was not being as useful as I might be because I did not understand the technology well enough. I asked for some teaching, and for a week was treated like an undergraduate learning about metal-cutting. Among other things I was shown a video which is used in teaching first-year students. It was an excellent teaching aid, but within the first five minutes two things had happened: (a) the operator had been referred to as a constraint, a cost. He was never mentioned again. And (b), the content itself, the engineering, was very fascinating and absorbing. These two things together would, of course, help to set a student's attitudes for life and be very difficult to counteract later.

Engineers and social scientists in the present age are therefore to a considerable extent products of this deeply institutionalised splitting. So powerful is it that large parts of both professions see no relevance in collaborating with the other at all. A feature of the situation is that some social scientists are afraid of technology and some engineers are afraid of getting into the human area. These fears are difficult to acknowledge and, from such fear, the human aspect may get turned into pseudo-mechanical form, like 'the Man–Machine Interface' or 'the Human Factor'. There is also a substantial history of mutual criticism.

On the side of the social sciences, more precisely of sociology, criticism has its roots in studies of the human and social consequences of production technology, and was originally not directed at engineering design but at the economic framework within which it was taking place. Marx's original analysis of the societal consequences of trends in manufacturing technology contained much of what a present-day social scientist would call socio-technical understanding; that is, understanding of the interplay between human and technical aspects of technology. The following is an example:

In the English letter-press printing trade, for example, there existed formerly a system, corresponding to that in manufactures and handicrafts, of advancing the apprentices from easy to more and more difficult work. They went through a course of teaching till they were finished printers. To be able to read and write was for every one of them a requirement of their trade. All this was changed by the printing machine. It employs two sorts of labourers, one grown up, tenters, the others, boys mostly from 11 to 17 years of age whose sole business is either to spread the sheets of paper under the machine, or to take from it the printed sheets . . . A great part of them cannot read, and they are, as a rule, utter savages and very extraordinary creatures. To qualify them for the work they have to do, they require no intellectual training; there is little room in it for skill, and less for judgement; their wages, though rather high for boys, do not increase proportionately as they grow up, and the majority of them cannot look for advancement to the better paid and more responsible post of machine minder, because while each machine has but one minder, it has at least two, and often four, boys attached to it. As soon as they get too old for such child's work, that is about 17 at the latest, they are discharged from the printing establishments. They become recruits of crime. Several attempts to procure them employment elsewhere, were rendered of no avail by their ignorance and brutality . . .[1]

However, although this is socio-technical analysis, Marx did not draw socio-technical conclusions, i.e. he did not conclude that social aspects should therefore feature in engineering design. He attributed the problems he saw to the ownership of private capital and the drive to create surplus value and did not take seriously, as an independent contributing factor, the simple need to reduce complexity and the resulting models of man in the minds of engineering designers; in other words, splitting. Since designers were generally working for the owners of capital the omission is understandable, and indeed there is some connection. But we know today that trends in design do not automatically change when ownership changes, as in nationalised industries or in socialist societies, or when the need for economy abates, i.e. during phases of subsidy. Splitting and its consequences are powerful independent contributing factors.

In the 1920s, thirties, and fifties a range of researches and other activities of social scientists began to elucidate specific rather than global problems. For example, empirical research showed that, given the opportunity, people varied their working pace in the course of the day, without loss of output.[2] This has never found its way into the kind of 'knowledge' that is explicitly used in design. Only last year, the managing director of a company in the domestic electrical appliance industry reported that he was 'amazed' at how miraculously their industrial relations improved when they took the mechanical drive off their assembly line.

What happened instead was a split development. Production engineers continued to work on the assumption that controllability and predictability required evenly spaced, i.e. mechanical, pacing. Then, in the 1950s, when basic standards of living were recovering from the war, the motor industry began to suffer from waves of strikes, most of which were unofficial and short. No-one recognised that the main function of a short strike is simply to create a break, an interruption from work, and that they were happening in situations where work was machine-paced.

Again, empirical research showed that, if peoples' actions were closely controlled as in work study systems, they would react by inserting controls of their own. Fiddling in work-studied incentive schemes had the function of exercising creativity and regaining control over one's work situation which the formal system did not permit.[3] And again, research showed that, if the work system did not provide feedback (knowledge of results), people would insert a way to obtain such feedback informally.[4]

The study which conceptualised much of this in a cumulative way was one which showed that, given experience of a job and some flexibility, people would find the optimum way of doing it for themselves. Conversely, if a new technology did not take account of their experience, its productivity potential was not realised: the technical system and the social system were truly interdependent.[5]

Some engineers and their institutions began to be interested in the findings of this kind of empirical social research. However, before a move towards integration in design could get very far, a second trend within the social sciences was making itself felt. The expansion of social science teaching and writing in the 1960s brought with it the re-emergence of critique as the dominant mode. This time it tended to be *mere* critique, on the basis of a pre-existing formal theoretical framework, rather than empirical and grounded investigation as had been the case before and was, indeed, the case with Marx. Given that the frame of reference for critique was well established, and frames of reference for synthesis and contribution only beginning to be worked out, critique was simply an easier option, and many social scientists chose it. The two trends – sociological critique and socio-technical design – to some extent came into conflict. The difference is a fundamental one.

At the time, public awareness of a need to bring social science and engineering together was growing. But it has turned out that, where arrangements have been made for social scientists to make a contribution to the education of engineers, they have tended to do it from a split position, i.e. they have tended to teach elements of social or psychological theory, not to help engineers incorporate human and social factors into their engineering. I personally regard this as one of the tragedies of the century.

There is little point in speculating about whether values developed as a result of the habit of critique, or the habit of critique was a consequence of values. It is in any case a fact, significant in the context of the UMIST project, that when we advertised for a social scientist to work with engineers in the UMIST team on the design of a technology, there were very few applications for the post, in spite of substantial unemployment among social science graduates. It was not the kind of job their teachers, in the late 1970s, had prepared them to want – or, indeed, to be able to tackle. Twenty-five years earlier, young social scientists would have given their eye-teeth for such an opportunity.

Engineers, in turn, insofar as they have been aware at all of what social scientists were doing, have resented being forever criticised. They notice that social scientists are not given to studying the ways in which human life has been made easier by the products of engineering. They consider that the social sciences have little in the way of positive contribution and find social scientists unwilling to try to help instead of criticising. And, if they experience resentment, engineers can get their own back. It is easy to trap a social scientist with questions which are not only unanswerable, but which serve to block the contribution that might be made. The following are two real examples, from experience. The first concerned the design of a new plant:

We want to design this plant so that the operators will be happy. What we need from you is advice on what colour to paint the walls to achieve this.

The second concerned the design of equipment:

Social scientist: 'We need to keep options open for the operator.'
Engineer (after doing a quick calculation): 'I reckon there are about four billion options. Which ones do you mean?'

This combination of recognising the value of the other and resentment of the other is the dynamic which psychologists call ambivalence. And both the habit of critique and the habit of resentment are sufficiently well established to have some of the characteristics of cultures, affecting to some extent even those, in both professions, who do wish to collaborate. It is against this background that attempts to work together take place.

Models of Science

As this discussion now moves on to the practical aspects of collaboration, it will confine itself to the concepts, methods and experience of those social scientists and those engineers who do wish to engage in collaborative work. For even where there is a wish to collaborate, there are still considerable handicaps. The phrase 'multi-disciplinary work' trips more easily from the tongue than it is realised in practice, and this does not apply only to these particular disciplines.

It is possible to postulate two models of science in approaching the topic of job design and work organisation: in one, there is a high value placed on measurement and quantification, in the search for precise guidelines. The other accepts ambiguity and conflict of interest as part of the reality being dealt with. (It is salutary to remember that 'scientific management' was to a large extent motivated by a wish to take conflict out of the situation by developing 'objective' standards which would be self-evidently correct and which would therefore be accepted by both management and workers. It has been one of the main causes of conflict ever since.)

It would certainly not be true to say that engineers necessarily adhere to the first model, and social scientists to the second. On the one hand, it was quite surprising to discover, in the course of the UMIST project, how much of engineering is still empirical (though that 'still' shows the power of stereotype) and how much room for debate there is among engineers. On the other hand, a good deal of social science research is conducted within the natural science model.

There are also cultural influences at play. In Germany, for example, there is a clause in the Company Law of 1972 which requires that 'proven scientific findings about the workplace must be applied'. This sounds strange to

British ears, where research involving people at work has been very context-specific and where the emphasis in application has been on cases and experiments rather than on the broad application of generalised 'knowledge'. The clause in Company Law in turn has had considerable influence both on the sponsorship and on the nature of research carried out since it was passed.

For example, a piece of social science research was used to test what level of buffer stock on an assembly line would be optimal in freeing the operator from machine-pacing. As part of a complex research programme a 'job satisfaction index' was compiled from the views which operators expressed about a range of things connected with their work. The job satisfaction index was then plotted against the buffers relating to their particular workplaces. A straight correlation was found. In other words, for whatever reason, the bigger the buffer, the greater was the job satisfaction. Far from indicating an optimal level, to which the law might then be applied, this brought the issue back into the arena of negotiation and the social scientists, no less than the engineers, were disconcerted.

The Disciplines in Relation to Values

There is considerable confusion about the connection between personal values, human needs, and the professional contribution to joint design. Social scientists do not have a monopoly on 'human values'. Nor do the social sciences represent some kind of authoritative statement about what human values are. What social scientists do have is, first, some understanding of what human beings need for their development. Values about job design and work organisation may then be expressed in terms of which of these needs has primacy. Secondly, they have some methods that help to articulate and make explicit the values and priorities held by people in a particular situation. And thirdly, they have some methods for operationalising these.

The first attempt to formulate needs to obtain meaning and self-development in the course of work into criteria for the design of work situations was made by Emery and Thorsrud.[6] While various attempts at catalogues of this kind have been put forward since then, and emphasis and interpretation may vary, the basic needs which are being articulated in that list remain constant: the need for autonomy, the need to use one's skills, the need for opportunities to learn and go on learning to limits set by oneself, the need for feedback as an aid to autonomy and learning, the need for purpose, and the need for interaction with others in purposeful activity.

In what follows, these will be discussed not in terms of priority but in terms of relation to the UMIST project. Purpose, in a flexible manufacturing system, is fairly clear. The meaning given to the term 'human-centred' in the project turned out to be in the areas of autonomy and the use of skill. The basis of the UMIST project was a recognition on the part of engineers that Taylorism had been (or rather, is) about minimising autonomy and the use of skill on the part of operators, and the central focus of the design activity was in these two areas. However, that left two more: interaction with others and development through learning. Interaction with others and group aspects cannot be designed into a piece of equipment, they are a matter of how that equipment is then installed and used by its purchaser. This problem, which concerns the boundaries of the system which the designer can influence, was discussed a number of times during the project but not resolved. We learned of a design group in Germany which entered into negotiation with potential purchasers about how their product was to be installed and used, but that does not seem a very realistic strategy. The best thing I could think of was something equivalent to a maintenance manual, designed not only to protect the features that are designed in (some of which could be counteracted in installation) but to extend them, on the lines of 'To use this equipment in the sense in which it was designed, you should . . .'

That leaves the need to go on developing through learning, and this was an area of controversy within the project. The engineers approached the 'human-centred design' task specifically from a wish to counter the effect of Taylorism, with which they were familiar, and this led them to emphasise autonomy and valuing the operator's use of skills, particularly tacit skills. The social scientists started, not from Taylorism specifically but from human development needs in general, and that included the importance of learning as well as the other factors. It was this difference that lay at the root of the 'blank table debate'.

The Disciplines in Relation to Outputs

As a group of professions, engineers have of course been evolving their methods and developing their products for much longer than social scientists. Social reflection about the output of technology is as old as technology itself (certainly the authors of *Genesis* had a view about the condition of man once he had to labour, and saw even God as needing a rest). But as disciplines which attempt to aggregate the outputs at a societal level or study them systematically at individual level, and thus to verify social reflection empirically and systematically, the social sciences are young by comparison. And as professions which attempt to contribute the resulting knowledge and methods in a variety of spheres and in the design of plant and equipment, very much younger still. As has been said, the very process of studying and commenting, by bringing gaps to light, has led them into critique and contributed to splitting, and thus hindered the development of the contribution.

Nor do the social sciences have much to show in the way of products, as products are seen by engineers. By definition, where an idea or a finding in social science is accepted as valid, it becomes incorporated into the general body of common sense. The idea of unconscious motivation is now part of the general view of human behaviour, featuring in literature, journalism and general conversation quite naturally, and without any need to refer explicitly to the body of psychoanalytic theory and practice which gave rise to it. Mothers are now encouraged to stay in hospital with their small children as a matter of common sense, not as a consequence of research which demonstrated the effects of not allowing it. Thus the useful products of social science are likely to be understandings, methods, practices and institutions. Sometimes they may be standards, but where standards are applied the processes of developing them, which social scientists may sometimes consider more valuable than the outcomes, because of the learning involved, will have been bypassed. Engineers, on the other hand, are likely to want outputs in engineering terms, i.e. at least in the form of standards.

Standards which social scientists can formulate with confidence and without an empirical 'research loop' are likely to concern processes, not outcomes. 'No installation without a transitional system involving prototyping', rather than 'no cycle time less than x seconds'; operators should have some say in shaping their environment, rather than 'the walls should be blue'. This kind of advice may be experienced as especially irritating when the particular population of operators does not yet exist.

The Disciplines in Relation to Methods

Where the social sciences are particularly strong at present is, first, in the concepts and methods for making values explicit and, secondly, in operationalising them.

If space is provided for this, it can in fact be a fundamental contribution. Given the long-standing and deep tradition of splitting, being 'human-centred' in attitude and assertion is not enough. Methodologies are needed for operationalising it; but this may not be what engineers expect. I have been surprised in more than one situation that people who would not dream of making decisions about materials, or temperatures, or surface finishes without some kind of systematic trials, insist on making or asking for decisions about people by some kind of inspired guesswork.

One is somehow reminded of how Italians describe the behaviour of foreign tourists in the face of Rome traffic. It is said that some foreigners, no doubt otherwise quite rational people, faced with the need to cross a busy street, put up one hand to stop the traffic, hold the other tightly over their eyes, and plunge. Like manufacturing design, it quite often works.

Take this example. In a shipping organisation, a number of policies including 'integrated crewing' were introduced, designed to encourage seafarers to identify with the particular employer rather than with seafaring as a whole. It seemed at least possible that people who go to sea do not want to identify with a particular organisation; but when questioned about why he was so sure that they would want to if given the opportunity, the manager concerned put his hand on his heart and said, 'because I feel it here'.

Another member of the Steering Committee, John Fox, made a similar point in a committee paper. He made it in the context of a discussion on intelligent knowledge-based systems (IKBS):

Few complete systems are ever evaluated objectively. Consequently wishful thinking about the virtues of particular computer systems is rampant. It is not uncommon for systems that are described as user-friendly to be rejected as unusable. Sometimes software design ideologies are formulated on the basis of fragmentary implementations or even pure intuition. The consequence of this lack of objectivity is that practical designers can't know what to believe – which IKBS ideas really are good ones ready for application and which are not. Part of the problem is the lack of an empirical attitude. AI workers do not seem to be expected to measure the performance of their techniques. Also there is a lack of training in evaluation methods ...

Secondly, the social sciences have a methodological concern for the links between process and outcomes. This has given rise to the concept of the 'double task'. Any organisation, team, committee, board, working party, has in effect two tasks: the primary task for which it was set up (sometimes referred to as task 1), and the task of designing, implementing, maintaining and reviewing the institutions, processes and mechanisms needed to carry it out (sometimes referred to as task 2). When task 1 is simple, or the people involved intuitively very skilled in these things, it may not be necessary to attend to task 2 in a very explicit way. Most often, however, it is necessary to deal with task 2 explicitly, and sometimes even to suspend business and pay attention to it, for a time, to the exclusion of other matters. The dynamics at the input end of design have direct consequences for the output end; in fact, that is often where the origins of poor design decisions are to be found. If design decisions are made for reasons other than design needs, the outcome is bound to be problematic. For example, a fierce argument about which of two layouts to adopt in a new plant was conducted on cost grounds (which were considered to be acceptable grounds for debate). It was in fact about the competition between the company's and its parent company's engineering departments (which was not acceptable). The costs could not be assessed in their own right unless the dynamics could be worked through.

'Working through' is a process of dealing with such issues which may be very hard work for those concerned, but which often leads to an apparently intractable problem being reformulated.

The fragment which follows is a small example of working through. Reference has been made to a difference in orientation between two views of what is meant by 'human-centred', with particular reference to the role of learning opportunities within that broad concept. In the project the difference had emerged as a debate about whether the operator should instruct the computer or whether the computer should make suggestions which the operator, from his experience, could override. Towards the end of the project an attempt was made to recapitulate that debate, with a view to trying to understand it and learn from it. The framework now was not conducting the argument, but working through the difference. Because of this, the problem got reformulated in the process.

Social scientist	Engineer
From the point of view of human development, there is a high value on learning	
	But why learn what the computer thinks it knows?
If we do it your way, there will be so many occasions when the operator accepts what he is given, that the assumption will be 'the computer knows'	
	If we do it your way, the operator is encouraged to enter a learning situation with the computer which is false

Social scientist	Engineer
The same is true if we do it your way. Also the operator will learn more and more what is in the computer and that will reinforce passivity. In my way he is making the *first* step	
	But his first step will increasingly be like the computer. He should be learning to laugh at the computer
My fear is that, since the computer will not be wrong all that often, and since designers are also learning, he is unlikely to reach that position	
	Your way would encourage him to concentrate on learning from the computer. My picture of you is that you think *all* learning situations are good.
I think he first needs to know how the computer does it before he has the confidence to disagree	
	He needs induction training on that, including that there are uncertainties. Something like 'in these situations be suspicious'

At this point the discussion changed to the nature of the induction training that would be needed. The reader should note the shift that has taken place:

1. There is now joint recognition that there is an area where two desirable objectives cannot both have priority and the nature of the mix needs to be clarified
2. There is joint recognition that the problem cannot be solved through the software design alone
3. The mode has changed from debate to separate contributions to joint problem definition

Since the boundary – the nature of the mix to be taught – was not clear, it was felt that operators should be invited to the discussion, that analogous situations should be sought and examined, and that practical examples should be included in the induction training. Induction training would also have the great advantage of introducing some human support and not relying on the man–machine interface for everything. Once formulated, it might be extended to a handbook.

In other words, once the framework shifts from debate to joint problem-solving, some of the other problems begin to be less intractable.

There then took place a redefinition of the central task, as it emerged from the need to formulate induction training and the need to clarify the relationship between the computer's knowledge and the operator's knowledge. It was to design the 'centre of gravity' for learning.

To illustrate how contact and joint work facilitate the internalising of concepts, it was now the engineer who formulated this as a *double task*, with two things having to go on simultaneously, thus:

Design the 'centre of gravity for learning'

Design the 'centre of gravity'   Facilitate the learning of the design team

---

Operational Issues

Phasing

Once engineers and social scientists are working together, many of the issues which have been discussed often emerge operationally as problems of phasing. Putting the disciplines and their concerns together does not automatically lead to integration, the consequences of the original splitting still have to be worked through. Otherwise, if a project or development process has been designed with the assumptions of one discipline, the contributions of the other may appear as things that will hold us up. The following are three examples of problems of phasing.

1. *Company A* was an oil company, engaged in building a new fuel-oil pipeline from one of its refineries to a major distribution terminal. Fuel-oil facilities would need to be built at the terminal, which until then had been engaged in the storage and distribution of other products.

One decision which had to be taken concerned the site of the control room. One alternative was to build a new control centre at the entrance to the site, so that truck drivers would be given instructions as they drove out, and hand in documents as they drove in. This would also have the effect of geographically separating the control of loading from the physical operation of loading. The alternative was to extend the existing loading and transport control room at the centre of the site. The engineer in charge of the construction came to ask what the difference would be in terms of the social organisation and attitudes of the drivers. It was an unusually perceptive question; finding an answer would involve doing some work with the drivers.

I estimated that I could have an answer for him in three weeks, but he could not wait. Major consequences for work roles and group relations were, of course, implied in the decision, and the engineer realised this. But he was locked into a schedule linking the design and construction of the new building with the opening of the pipeline and could not create a three-week delay. He had assumed that an answer might be available ready-made.

2. *Company B* was engaged in building a new high-speed canning plant. It was to be built on a site where the company already had some other

operations so that, while the operators who would be manning the plant were not yet available, there was a trade union organisation. A Job Design Committee was established for a time, including representatives from three unions, to consider the nature of the jobs being created in the cannery.

Two control systems engineers were involved in the planning of the cannery. They became interested in the idea of job design and, after some preliminary induction to the topic, one of them gave a presentation to the Job Design Committee. He said that, at that stage, the control systems could still be designed in almost any way the Committee wanted. He liked the idea of working as a service to the operators who would later be doing the jobs. But he pointed out that, once the floors were laid, with channels for the cables, it would be very difficult to change. The trade union representatives at that stage did not have enough knowledge of the process to be able to be very specific about what they would want. The technology was very new and advanced, and the engineers had about two years' start on them in thinking and learning.

The problem of participants, or social science professionals, being out of phase with engineering designers in their absorption of the necessary know-how is very general.[7] Unless arrangements can be made for it to be dealt with, contributions are likely to be limited to general-level statements and 'participating' is likely to be superficial and unreal.

3. *Company C* was also building a new plant, this time for the manufacture of confectionery. Much of the production machinery was to be transferred from their existing older building, and the job design contribution concerned the organisation of work around existing equipment rather than the design of new equipment.

When I first met the company's Project Group, a site for the new factory had been acquired and planning permission obtained, and they were beginning to discuss architects and the general shape of the building. Within ten minutes of joining one of their meetings for the first time, I discovered the problem of phasing. Although I had been involved as early in the process as one may reasonably expect, from some points of view it was already too late.

The prospect of an entirely new factory, an opportunity which people have only very rarely, was acting as a focus for a powerful vein of idealism in the company. Not only did they want the jobs in the new factory to be satisfying for the people working there, they wanted the architecture to be innovative, to be human in scale, and to make a distinct contribution to the built environment. This had involved a good deal of discussion about company philosophy. Two concepts for the new factory were being debated: on the one hand the concept of a large, hangar-like structure, within which there would be freedom and flexibility to arrange and re-arrange things; and on the other hand the concept of a village street, with small production units, as well as social facilities such as a tea bar, a bank, possibly one or two shops, adding up to give the 'feel' of a varied and small-scale, village-like environment, where people moving from one unit to another would inevitably meet each other. This would be far removed from the conventional idea of a factory.

Within a few minutes of joining the group I was confronted with the question: 'What do you think – large hangar or village street?' I had, of course, no basis for an opinion and realised the dilemma we were in. The concept with which I intended to work was that of the production system as a socio-technical system. To translate this concept into practical reality, one needs to understand the manufacturing process and its technology in some detail.

The company felt that they could not even begin to talk to architects until they had some idea of the basic shape of the building they wanted; one could not sensibly discuss the shape of the building without some idea of the production layout; and I could not contribute to discussion about the layout from the job design point of view without some socio-technical analysis of the production process. At that stage I had not even seen the production process.

Two meetings and some familiarisation later, we achieved a breakthrough. I was still far from really understanding the details of the production system, but I had learned that the first of the products to be manufactured on the new site consisted almost entirely of crushed sugar, with some additives, which was then compressed into a tablet and packaged. Groping for a more detailed understanding, I said, 'Look, I still haven't understood the process properly – suppose I'm a piece of sugar, I've just been delivered. What happens to me?' Somebody said, 'Well, the first thing that happens to you is that you get blown along a tube. But there is a physical limit to how far you can be blown.' I said, 'O.K.; what happens next?' And somebody said, 'Next, you get crushed into a powder'.

In this way I talked my way through the process in great detail, role-playing the product. For example, I heard myself saying, 'All right, so now I'm a granule. What happens next?'

'Next, we drop mint oil on your head'.
'Might you miss?'.
'Yes, we might'.
'How would that be discovered?'.

And so on.

I checked back a number of times to ask whether this was just a game or whether it was useful, but they assured me that they were finding it very useful. The product was a fairly simple one, which they had been making for a long time, and their ways of thinking about it had become rather set. Now, these ways of thinking began to unfreeze, and they began to discover alternatives and to say to each other, 'It doesn't have to be like that, it could be like this, if such-and-such conditions are met'.

In particular, some things which they had been used to thinking of in sequence could, it was found, be done in parallel. That meant that the logic of the production process was not necessarily a straight line and this, in turn, meant that one could think in terms of a short, squat building. This was the eventual shape of the 'product house' which emerged out of this process.

In terms of work design, I realised afterwards that my strategy had been about leaving options open. Once the factory was staffed and experience of the work system beginning to accumulate, there was more chance of reviewing and revising it in a short, squat building than in one where the logic of the layout led to long, straight lines. In these, more things would be irrevocably fixed. From the socio-technical point of view, the short, squat 'product house' should have three functions: the opportunity to identify with the product, the opportunity for people to relate to each other, and the opportunity to keep design and organisational options open.

Architecturally, the 'product house' turned out to be a kind of compromise between the 'aircraft hangar' and the 'village street' ideas, smaller than a hangar but larger than the cottages envisaged along the village street. With this concept the company then went in search of architects.

Systems Boundaries – The Meaning of 'Human-Centred'

Within the human and social sciences there is a long-standing debate about how criteria which make a piece of equipment easy to use relate to criteria which make the work being done meaningful and which are conducive to the development of the person doing the work. In some instances they are identical, e.g. safety and health. In some instances they are different, e.g. the operator's autonomy does not generally feature among the usability criteria which 'Human Factors' professionals apply; and in some instances they can even be in opposition, e.g. usability criteria usually include short learning times, while developmental criteria give a high value to opportunities for learning.

These differences are sometimes debated as differences in values, but may also be discussed as differences about systems boundaries, i.e. about what is considered to be within the system, and therefore susceptible to design and change, and what is outside the system and part of the environment, i.e. to be taken as given. The issue is likely to crop up within multi-disciplinary design teams as well. It has two facets: on the one hand, how far responsibility goes for the impact that a technology will have and, on the other, what parameters to include in design considerations.

It may be useful to think of this problem in terms of a typology of products. Where a product is a *tool* as a means of doing something else, or creates a *task* which is only a part of a role so that the configuration of the role itself is out of the hands of the product designer, the aims of good integrated design may properly be in the direction of 'usable'. Where a product or system creates *roles*, and where the configuration of the role is therefore in the hands of the product or system designer, usability should be a minimum baseline, and the aim of good integrated design must be more in the direction of 'developmental'. At any rate, in such a situation the issue cannot be ignored. The product or systems designer cannot help influencing long-term consequences such as future personal development or organisation, or industrial relations, whether he wants to or not.

One can in fact postulate more than one such scale, and the boundary will not be in the same place on all of them. It may be useful to draw a profile across them for a particular product, such as:

From	To
Affects task	Affects role
Tool	System

Where a product profile is near to the left-hand column, it is not unreasonable to restrict 'human-centredness' to usability criteria. The nearer it is to the right, the more do developmental criteria need to feature as well. This approach would by no means eliminate value-based debates, for instance on the question of when to treat people as responsible in matters of safety and when to eliminate 'the human factor'; but it may help to contain them.

Making It Happen – Institutions and Infrastructure

Where the idea of the collaboration between social scientists and engineers is new to people, they are likely to become preoccupied with what this will actually mean in practice and how to set about it; they are much less likely to worry about longer-term considerations. However, this kind of work has now been going on for long enough for some long-term issues to become clear – not so much how to make it start, as how to make it stick. The issue is institutionalisation.

To institutionalise something is to build it in. It has been pointed out at the beginning of this chapter that the problem is not so much that the technical and social aspects of engineering design have become split off from each other, as that the split is deeply institutionalised.

The general idea may be illustrated by an example. Society has made a decision, crystallised in law, to restrict driving to one side of the road. This apparently simple decision is supported by a surprising number and range of institutions: the assumption that it must happen is built into the *design* of vehicles. It is built into the *training* of drivers, as well as into their *legitimisation* (licensing). It is built into the formulation of *codes* and *standards* (the Highway Code, standards about the width and layout of roads, etc.). Then there is the *continual reinforcement* of seeing that others do it and, finally, *sanctions* (punishment) if it does not happen. These institutions, in turn, are supported by *funds, training establishments, staffing* and *monitoring* (traffic police).

It is the combination that makes these institutions, together, powerful and effective. In addition, a breach of the decision is generally clearly visible and unambiguous. As a result of all that, the decision is mostly carried out: drivers are not continually deciding on which side of the road to drive.

Social aspects of work systems are, of course, not often so unambiguous, and some of these institutions would be inappropriate. But it must be remembered that this is partly a matter of local tradition. In Germany, as we have seen, it is the law, with its accompanying sanctions, that says that 'scientific findings about the workplace must be applied'. In Germany, too, human factors methods and findings form part of the substance of some trade union agreements.

Where something can be formulated as a standard, that is a form of institutionalisation. Where it can be integrated into the technology, such as dialogues that provide genuine options, or indicators on a machine that provide feedback, or buffers that mitigate pacing, that is a more powerful form.

But to ensure that such structural influences are taken seriously in the first place, and for situations that cannot be formulated in these ways, the need is to achieve a degree of culture change. To attain this, it is not a matter of one or more social scientists joining a team; the socio-technical viewpoint must be represented powerfully enough, early enough, and consistently as a matter of routine. It must feature in the syllabuses of engineering students, in the appropriation for capital investments, in R & D budgets, in the qualifications and experience required of systems designers and in the assessment and emoluments of technical directors.

It has taken me a long time to recognise what a mistake it is to appear to be solo when going into this activity, rather than institutional. First, a social scientist working solo as a member of an engineering activity is likely to be at the receiving end of all the ambivalence, the hopes and anxieties and resentments, that have accumulated in relation to the social sciences. If the things he or she says are unwelcome, they are likely to be seen as personal rather than professional contributions. Secondly, he or she will not have at his/her fingertips all the substantive and methodological contributions of all the social sciences. And thirdly, the number of things he or she has time to deal with will be very limited.

This is so even within a small team working on the development of a piece of equipment. It is much more so in the design and development of large-scale plant, where there are so many things going on at the same time that it is hopelessly optimistic to see them all as consistent and rational tributaries of the mainstream. So, for example:

There may be market changes during the design and building process which influence the production capacity that is needed, leading to decisions outside the design teams.

Equipment will be purchased off-the-shelf and therefore cannot be influenced unless this influence is exerted long before actual orders are placed, i.e. unless the culture of the equipment manufacturers can be influenced as well.

Engineers move out of teams and are replaced by others who have not shared the history of collaboration. The social scientist is likely to have put the main effort of developing collaboration in at the beginning and may not recognise the need, or have the time, to start again from scratch.

All these instances illustrate that the system which is carrying the design and development processes is an open and not a closed one. In these circumstances, mere collaboration between social scientists and engineers is not enough. Mere collaboration, however well it may be working, is too weak a mechanism to cope with open systems characteristics. On the other hand, mere infrastructure is not enough either, and formal guidelines circumvent the necessary processes of development and mutual learning. What is needed is collaboration supported by institutions and infrastructure.

References

1. Karl Marx, Capital, 1958, vol. 1, pp. 484–485 (Lawrence and Wishart).
2. D.W. Harding, A note on the sub-division of assembly work, Journal of the National Institute of Industrial Psychology, 1931, vol. 5, pp. 261–264.

3. Lisl Klein, Multiproducts Ltd: A case study on the social effects of rationalised production, 1964 (HMSO).
4. S. Wyatt and J.A. Fraser, assisted by F.G.L. Stock, The comparative effects of variety and uniformity in work, Industrial Fatigue Research Board report no. 52, 1928 (HMSO).
5. E.L. Trist, G.W. Higgin, H. Murray and A.B. Pollock, Organisational choice: capabilities of groups at the coal face under changing technologies, 1963 (Tavistock Publications).
6. F.E. Emery and E. Thorsrud, Form and content in industrial democracy; some experiences from Norway and other European countries, 1974 (Tavistock Publications).
7. Kenneth Eason, The process of introducing information technology, in B. Shackel (editor), Usability in human–computer systems, 1986 (Addison-Wesley Pub. Ltd).

Chapter 7

(How) Can Technology be Redirected?
A Scandinavian Perspective

Håkon Finne

Introduction

Is it really possible to redirect technology, and if so, what does it take to do it? These were among the questions that were on my mind when I first joined with the project; and, while I still feel there is no definite answer, I would like to share some reflections from the outside on what it might take and how I would interpret the project in that respect.

My connection with the project stems from a six months' residence with the group in 1982, i.e. while the research team was still being recruited.[1] Apart from a shorter visit and personal contacts with some of the participants, my insight into the later stages of the project has been limited to that which I could read from discussion papers and minutes (as well as some tape recordings) from Steering Committee meetings. My commitment to its cause stems from my own research experience over the last ten years in trying to create a meeting ground for social science and technology and working life in order to improve working conditions in a wide sense of the word. And while my interpretation carries much ambiguity as to the possible success of our efforts, this is not a concession to all the devastating critics in both disciplines that I have met in my time, but rather a caution to the adherents of disciplinary integration to look beyond the disciplines themselves.

My approach will be as follows. First I will consider how one could envisage design criteria for the more tangible research result, i.e. software for an FMS in which operators are not subordinate to machines. More precisely, I shall comment on how this aspect was conceived in the project and also offer my own understanding of it. I will then examine the effectiveness of projects like the present one on two levels. First, I try to establish the effectiveness of the chosen methodology in achieving the main objectives. Secondly, I take a look at the efficacy of setting up a project like this given that a redirection of technology is its raison d'être. Finally, I try to address the question of further development.

The Concept of Non-subordinating Machinery

Surprisingly early in the project, Martin Corbett operationalised (on the basis of social psychology) the quality of non-subordination into a small number of design criteria which were well received by the Steering Committee. Some of these were of course traded off towards other criteria in the course of the design process. For example, one of the main design criteria was to minimise the shock to the operator. This might call for a congnitive congruence between the optimisation strategies of the computer model (for selecting cutting parameters) and the worker. However, the project found that under constrained conditions, some workers will reduce speed before feed while others will do the opposite. The project decided to include only one of these strategies in the software, since leaving the option open for customising would contribute to raising the cost (or computer requirements) for the system.

As the work progressed, however, it became increasingly clear that despite the agreement on Corbett's design criteria, there was an underlying difference in the conceptions of control and subordination. This difference was highlighted by means of the 'blank table debate' and by the emergence of the competing concepts of non-subordination and human-centredness. As far as the whole system goes, there clearly is agreement within the research group that the software gives more space for skilled operator intervention than does the traditional NC programming approach, and I share this view. The disagreement concerns whether that space is sufficient, and, in particular, how the system can provide learning opportunities that will allow the operator to remain in control.

Note that there is a seeming agreement in locating the control issue to the working situation, including the prospect for learning how to increase operator control over the metal-cutting process. While this constraint seems appropriate because the project aims at producing workplace technology, my personal feeling is that the debate could be enhanced by understanding control in a wider sense.

The difference in position has been explicated in terms of the sun-and-corona metaphor and explained by value differences. While I find that analysis intriguing, I think it could benefit from a clarification of the relation between the sun and the corona. More specifically, I shall consider the relation between tacit and scientific knowledge, then link this to learning, and eventually relate that again to various concepts of control.

A strong premise of the project is that tacit knowledge and scientific knowledge are of different kinds and that they have equal or similar importance for production of material goods, economic value and human experience. Little has been said explicitly in this book about the nature of these two, however, and I shall try to convey my understanding of it.

Scientific knowledge about metal cutting includes quantitative relations between various measurable properties of the cutting conditions. It excludes variables that cannot be linked to these conditions by a causal theory (or scientifically plausible hypotheses). Whether expressed in formulae or tables, the validity of this knowledge appears as constant over different contexts, except that the context may not allow its full application. As a corollary, the formulae appear without any connection to their mode of original production. This knowledge can be applied by anyone who has internalised a technical outlook and has learnt a number of manipulation techniques tailored to the task, and it can be incorporated into a computer program.

As an example, let us have a closer look at the cutting technology. Software support for the selection of cutting parameters was based on the extended Taylor formula. The original formula was developed by Frederick Winslow Taylor[2] and related a number of parameters of metal cutting to each other with the purpose of reducing cost (i.e. cutting time). Taylor spent several years accumulating empirical data for this. His laboratory was a commercial steel plant with piece-rate skilled workers and machine tools. The original reason for beginning these experiments was, according to Taylor's own account, to objectify the tacit knowledge of the skilled metalworkers in order to break their 'soldiering' practice of collectively restricting output, in particular when faced with cutting tools of unprecedented hardness. Several and severe disciplinary measures were required to establish the laboratory conditions necessary for producing scientifically valid data, as the workers were not interested in disclosing their knowledge to management, fearing (with reason) that it might result in reduced piece-rates.

After those trail-blazing experiments, much effort has gone into refining Taylor's models and obtaining new data as new materials and cutting techniques became available. Astonishingly few attempts at automating the selection of cutting data have been successful, however, reflecting the difficulty of the task.[3]

Once extracted and expressed, however, this scientific knowledge bears little mark of the human conditions of its production. It can be used for many purposes, including educating metalworkers. It can also be used by management, or technical specialists, or computer programs, to specify cutting data and hence remove the control entirely from the worker. Or the skilled worker's refinement on these specifications can be elicited, as in the present project. Because of the complexity of the calculations, computer support can give more time savings (on cutting) than the calculations yielded by the simpler slide rules and tables which were conservative implementations of the same knowledge.

Let us now return to the other form of knowledge. Tacit knowledge is essentially that which you know but cannot tell. Theoretical scientists, business executives, and metalworkers all develop tacit knowledge relating to their professions.[4] While some tacit knowledge may be formulated, other parts are believed to evade extraction even in principle. It can only be inferred indirectly as a prerequisite for skilful behaviour. This is partly related to the issue of hand and brain, to which I shall give some consideration.

Physical interaction with force and tactile feedback were believed to be important or even essential in learning to become a good metalworker. Interposing a computer between the physical controls and the person would remove one of the crucial links, as Cooley has noted (see Chapter 10, p.138).

The conceptual difficulty embedded here is shown by an incident early in the project, where one of those attending a Steering Committee meeting made an earnest but erroneous attempt at substituting hand-and-eye coordination for the hand-and-brain issue. While hand-and-eye coordination involves only sensorimotor skills, the hand-and-brain issue emphasises the links between sensorimotor abilities and mental forms of knowledge.

There exists some strong evidence of the importance of this relationship. Developmental psychology definitely shows the importance of bodily exploration of the physical environment in children's intellectual growth.[5] Some branches of research into artificial intelligence also consider the 'physical intellect' to be the basis for intelligent behaviour in the world.[6]

Less is known about development of metal-working skills in adults, however. As mentioned before, no previous research was found that could resolve the issue. One scientist considered the link as potentially important for personal development but obsolete in terms of learning metal cutting, hence he suggested more contact with nature in one's spare time. Another scientist claimed that many people showed a quicker learning curve and performed more rapidly and reliably when dealing with spatial locations of, say, control panel instruments than with symbolic strings of the same data.[7] This might possibly indicate an analogue hypothesis for the influence of tactile feedback on personal performance in metal cutting.

While the project members all seemed to share the idea that physical contact is required for learning, the computer was not paralleled by a more direct device because of the assumption that this kind of training would have to take place outside an FMS setting as a basic acquisition of metal-cutting skill. It is still unclear, however, what role the physical interaction would have in refining skills beyond the capability of the automated cutting parameter selection.

I have devoted much space to physically acquired skills. It is important to note, however, that these are not the only ones that are covered by the concept of tacit knowledge. I shall now consider another important aspect of this which is termed theory of action.

A theory of action is a person's set of norms, strategies, and assumptions about the (causal) relations between the strategies and the norms under various conditions.[8] Outsiders wishing to establish this set have to do so empirically by observing the person in action, since the individual may espouse a different set without being aware of it. Usually, the theory-in-use is more sophisticated (but sometimes less consistent) than the person's espoused theory.

This theory-in-use is largely tacit, and it derives from (and belongs to) the person's personal knowledge. This knowledge consists of facts, beliefs, and assumptions, science-based or not. Much of this is connected to the setting in which it was acquired, i.e. it has psychological, or social, or political, or economic, or affective connotations. (Scientific knowledge is presumably without such connotations, but scientific facts learnt by a person have them.) It is definitely connected to purpose, or else meaningless. Some of this knowledge is acquired by means of all five senses and becomes embodied in a literal sense, particularly that which concerns activities we are good at (e.g. everyday life activities and our professional skills).

The way we value knowledge of different kinds as persons depends to a large degree on the ethos of the learning situation. This valuation process to some extent influences our theories of action, as will be illustrated by comparing the teaching of metal cutting in a craft ethos and a scientific ethos.

Metalwork used to be taught by practitioners and the instruction was void of scientific content. Metalwork was learnt by working with a number of experienced practitioners, and relating this to accumulating personal experience. The theory of action would constitute local contingencies as the norm, standardised conditions as the exception. Development of one's own strategy was fundamentally a social process. Novices were strongly urged to stay in line with the local practice. Skilled turners would relate strategies of others (comparing their strengths and weaknesses) to each other and develop their own. Hence some local cultures would reduce feed first while others would sacrifice speed, and some might make that contingent on how appropriate they found the designer's choice of surface finish and the likelihood that he would ever learn about the deviation on this particular batch.

Scientific metal cutting would deviate from the craft ethos on a number of important points. With a science-based education, the curriculum comes closer to advocating a one best way, with local variations as unfortunate deviations from the norm. The expertise would tend to be based on a single, unified body of knowledge resting with a single source of authority.

This ethos must not be confused with that found in learning how to practice science. The conduct of science relies heavily on the tacit knowledge of its practitioners.[9] In this sense, science is very much like a craft. In its external workings, however, science discounts any knowledge that cannot be formulated as explained above. Scientific knowledge is transmitted to non-scientists as definitive, in a way that strongly implies a division of labour where scientists create knowledge while others comply with it.

I will maintain that an education with a scientific ethos would, if effective, make it more difficult for its students to lend ear to non-scientific learning experiences. Hence their theory of action would be fairly close to the causal theories of metal-cutting science. They would of course learn that 'in practice, you cannot live up to theory', but I believe the scientifically dominated theory of action would effectively bar their further development into highly skilled operators if it was maintained by a scientific ethos in the workplace.

On the other hand, I think it would be easier to incorporate scientifically obtained data and theories into the education were it to be craft-based. In this ethos, science-based facts would be one of many kinds to be integrated into personal knowledge. After some years, then, a turner's theory of action might include something like the following: 'On rush orders for conical axles the Taylor slide-rule has proved to be good, but if their holes must be drilled on Jack's old machine, then I'll use the procedure that Johnny and I worked out over tea that late shift in '86 when the foreman would explode every time Johnny requested a new drill bit.'

Now this discussion of craft ethos versus science ethos would be entirely external to the project had it not been for two facts. The first is that the project assumed primary training to be a prerequisite and the second that the FMS technology could contribute to setting the ethos in the workplace. The second factor becomes even more important when we consider that not all roles in the FMS will be filled with fully trained workers and hence the skilling process of junior operators would be strongly influenced.

My own analysis – which is admittedly very tentative – deviates from that of the project. While they were trying to figure out what the blank table could contribute to learning in itself, I suggest that this depends strongly on the ethos. I hypothesise that in a learning climate coloured by a scientific ethos, the blank table is of little use except as an instrument for testing one's skill against that of the computer (or the computer programmers, rather) in the area where the computer is at its best. In a craft-ethos learning environment the blank table may be necessary, or at least very helpful, for learning. The craft ethos might allow a multitude of ways for using the blank table (and the rest of the software support), as well as multiple ways of interpreting the results.

While none of the interviewed workers saw any need for the blank table, I think this response evolved from the lack of experience with the cutting technology. They had not yet experienced any area where the calculations were in strong disagreement with their own preferences. Equally important, the metal cutting itself was not an integral part of their evaluation. Hence the core of the labour process was external to the learning simulation and indeed seems to have been external to the debate on the merits of the blank table itself.

There is of course no guarantee that the blank table will be used in a specific way. Time pressure and output norms in the working situation often tend to make experimenting impossible, as my own data from Norwegian firms suggest, even if there is explicit top management support for operator learning. However, German experience suggests that CNC technology, originally intended for large firms with a high degree of division of labour, finds its more innovative application in small firms with skilled workers. This was not so with the less flexible NC systems, and the blank table would support learning experiments better than the editing facility alone, even if we cannot say now exactly how.

With a scientific ethos, science holds the authority of knowledge, and there would be less encouragement or invitation to put the algorithmically calculated data to an unscientific test. (The purpose of in-factory testing as part of daily operations could not be scientific, of course.) In this situation, the blank table could be seen only as an instrument for internalising the scientific data as well as their legitimate validity. This, I think, is not what the project aimed at.

Returning now from the learning issue to our original concern, what difference does the ethos of the learning situation make for the control aspect? In terms of the daily control over the metal cutting, I think the difference is minimal in most cases. The default cutting values will roughly determine output norms, and the set-up (with rapid manual geometry input, automated cutting technology, editing facilities, and a blank table) will leave the marginal improvements with the skilled operator and not with programmers in the office.

If we are thinking about the long term, however, my answer is different. This is more based on an intuitive guess which I will explain below than on empirical evidence. In one case, the daily working experience would continually verify that there was only one source of accurate knowledge (science), and it had only one form (explicit). While metal cutting in practice sometimes had to deviate from this standard, that would be (considered) a flaw in the operations of the machine and not in the nature of the scientific knowledge. In the other case, the daily experience would be one of continually confirming that personal, tacit knowing and explicit scientific knowledge might complement each other, and that there was more than one source of valid knowledge. In other words, the consciousness produced in the two instances would be different.

Either of these cases might be reinforced or weakened by experiences in other fields of life, of course, but I definitely see some conflict between the essentially authoritarian reliance on only one form and source of knowledge and democratic values. In this interpretation, the scientific ethos would mean that the metalworkers' control over the metal cutting would not be a personal control. Rather, the individuals would be the extended arms of a control that resided in the scientific community.

The sun-and-corona metaphor related the argument back to differences in values. My analysis also relates to values, although I do not claim that one party to the argument holds scientific knowledge to be superior in an authoritarian way. On the contrary, the value placed on both kinds of knowledge is remarkably evident. What I have in mind is more a value that I would ascribe to science as an undertaking given its position in society. In my opinion the possible links between scientific progress and authoritarian values have not been properly addressed yet. It has always been assumed that the original achievements of modern science in liberating mankind from dogmatic authority would guarantee the liberating effects of educating people to trust in science alone for their knowledge needs. That may not be altogether true any longer. Perhaps it is this question that is tacitly nagging those who wish to maintain obsolete skills because they are important to personal development. And even if it is not, I still think it is worth considering.

Methodologies for Designing Redirected Technology

In this section, I shall discuss the methodology chosen to reach the two main goals and some possible alternatives. At the time the project was designed, I had no better suggestions myself; all that is written below comes with the advantage of hindsight.

To recapitulate, the project design involved a multi-disciplinary Steering Committee that would develop a strategy for the project and a cross-disciplinary research team. (By multi-disciplinary I mean that many disciplines are represented; cross-disciplinary involves performing work that would synthesise knowledge from several disciplines or explore the empty space between them.) The social scientist, the engineer and the computer scientist on the team had unclear roles from the beginning of the project as is easily understood. Since engineering design research was presumed, however, there was more debate about the role of social science than that of engineering in the design exercise (although of course the *content* of engineering decisions was continually challenged). One major role of the social scientist was, however, to represent the knowledge of the context of prospective use of the system to be designed, i.e. to input knowledge about workers and work.

Social science is bound in an eternal dilemma in design situations. On one hand, social scientists will claim to know more about relevant social systems than the engineers on the design team. This claim will be made on a scientific basis. On the other hand, part of this knowledge includes knowing that a context (of, say, work in a particular firm with a particular set of machinery) can only be fully known from the inside. Hence they will not guarantee the positive status of their substantive knowledge, not because it may be wrong, but because it intrinsically cannot be either right or wrong, only context specific. Hence worker participation is often induced to overcome this problem, and the contribution of social science is partly one of facilitating the participative design process and partly one of bringing in substantive knowledge from other contexts.

Of course the research context is not the use context and so neither workers nor researchers can know the future context from within. It then becomes a question of transferability of knowledge from one context to another and, of course, the possibility of simulating the future context.

This problem was addressed in another way in the project. The question of worker participation on the research team was raised but dismissed for compelling practical reasons. Instead the issue was shifted to possible worker and/or trade union representation on the Steering Committee, and to the question of research on worker reactions to the proposed and similar machinery. Hence conventional CNC users were interviewed and workers were invited to study a prototype of the software and were interviewed about this and about possible future features of the software.

While there was agreement that data thus obtained were by no means sufficient, no better option was seen. Experience elsewhere does, however, point to alternatives. People easily bring their explicit and tacit knowledge about contexts with which they are familiar with them into other settings. Hence workers can relate new technology to old experience provided they have a reasonably good feel of what the new thing is going to be like. But in order to contribute extensively to the design, they must be able to experiment with it. Various laboratory methods have been devised that would make it feasible to simulate the physical and functional properties of new technology. In particular, rapid testing of ideas can be done by using mock-ups. Workers would then work with the realised parts of the system, and, whenever an object or a function not available was required, it would be faked.

Hence in one case, a cardboard box was used to represent a laser writer for several months (before laser printers were readily available in their present form), and its specifications were changed many times over on the basis of recording what uses it was put to in the laboratory.[10] Another method particularly apt for software development is a group of techniques labelled rapid prototyping or experimental systems development as also discussed by Holden (see Chapter 5, p.74–75). With these techniques, specifications can be changed many times over on the basis of preliminary testing of incomplete prototypes of the software under development.

Similar conclusions derive from a study of industrial flexible automation development projects. Working groups consisting of designers (technologists), experienced workers and (in those cases) ergonomists, consistently produced not only the best designs, but also those that were best accepted.[11] The success was ascribed to the multiple competence of the design group (with overlapping knowledge about the immediate design task accounting for the ease of communication) and to the working methods, which included design on paper and work on physical prototypes in a laboratory-like setting.

There would have been few obstacles I believe to including the use of mock-ups and physical models in the experiments. This could have enhanced the possibility of relating working experience to the new developments. Of course the lack of access to the actual lathe would be a drawback anyway. Also, the funds were insufficient for more worker input. Experience shows that these artefacts are useful only for longer term commitments in the laboratory.

On the question of experimental prototyping, it would probably be a project in its own right to develop a suitable prototyping language for CNC software development. This would be an interesting approach. Many of the options closed during the project might have been kept open without too much cost attached to them had the project been about developing an application generator (or a prototyping environment) for CNC software rather than an actual system. The design specification would then rely first on identifying as many as possible of the choices that might be addressed in future applications. Many alternatives then could have been simulated at the next stage and evaluated in a more interactive manner. That would have made for an entirely different project, though, with necessary rethinking of the role of social science input.

Before endorsing this approach, however, a few comments seem necessary about previous experience with prototyping as a tool for worker participation. Current prototyping methods for administrative systems do not seem to afford the promised user control over development as often as hoped for. One reaction often found is that prototyping does not put the user in the position to develop better specifications than the expert.[12] Another reaction is that there is much more to participation than just selecting between the options usually involved in prototyping. This may partly be due to the prototyping tool but also to the context of application.[13] Social science input could be useful at both stages.

Still another approach would seem an appropriate candidate for a project like this, as a complement to what has already been mentioned. In order to specify the design criteria for the FMS, the team relied on two sources. One was to identify assumptions in the main thrust of technology development and to negate them whenever appropriate. Complementing this 'negative' source was a 'positive' one, i.e. research on what would constitute operational definitions of non-subordination. In both cases, however, analyses and categories for analysis were developed by researchers, and their validity would later have to be checked by putting the final system to work.

One could instead envisage using a method called a search conference in identifying issues to be attacked.[14] A search conference involves grouping together a number of people who will discuss their own future. Local communities, industrial firms, research customers, voluntary organisations, government bodies, school communities, and others, have used this approach to go through a sequence of activities in which they identify positive and negative historical trends affecting their future as a group, explicate the direction they want to move in, generate plans and ideas on how to get there, and agree on affirmative action. The event is run as an alternating sequence of group work and plenary exchange sessions over two to three days. The selection criteria for participation as well as the composition of the initial groups are strongly influenced by the organisers, but after that, the specific topics of each group and the individual group affiliation are determined by the participants themselves. Only the general drift from history towards affirmative action for the future is mandatory. With this method, both the categories of analysis and the suggested directions of development are rooted in the local knowledge structure and social reality of the participants.

While this approach has worked wonders when the plans are to be carried out by the participants themselves, there are as yet few experiences as to how, say, technology research projects could benefit from such conferences. In some instances, the researchers initially have found their own thinking to be more sophisticated about the technological matters while the workers and managers have been caught up in daily trivia. Further analysis has revealed, however, that an approach that was related to trivia of this kind would be necessary for developing a climate of acceptance of, and interest in, the proposed technology.[15]

In our case, a search conference might tentatively have been composed of skilled NC and CNC users in small firms, some unskilled (or semi-skilled) labourers who were integral to the production process, some NC programmers and process planners, some managers with executive power, and some researchers who would be developing technology. These would be people whose fates would rely on each other to some extent. The conference then possibly might have to be run by someone disconnected from, i.e. with no interest in, the project. Typical outcomes would include some that the project could deal with and many that other participants would be the right ones to handle; but the entire experience could create a social (and cognitive) space in which both the technology designers and the designed-for were committed.

How far into redirectional efforts can these alternative methods be carried? Most of them have been used for marginal changes compared to a redirectional effort, and they certainly should not be expected to replace the researcher-driven generation of a new trajectory. Nor can they solve the problem of long-term validation which was sought in the project. They would, however, contribute to extending the use context into the laboratory, which I would consider an advantage.

Was the Project Appropriate and Efficacious in a Redirection Effort?

Up to this point I have discussed the researched system and the paths that led up to it, comparing with other possible paths and similar systems. Now I would like to ask whether the project has been efficacious in redirecting technology, or more accurately, what role the making of an alternative FMS and a new design methodology could possibly play towards that overall purpose. Cooley has already presented a view of the project as a building block in a larger undertaking, including the significant ESPRIT project (see Chapter 10); my approach will be more analytical and try to hint at some criteria for such building blocks.

While the malleability of technology and the relative autonomy of technological development by now are largely accepted in principle, few attempts have been made at stating a coherent theory for delineating the conditions of a redirection effort.

I shall borrow a conceptual apparatus from the study of directional changes in science and see if it may be applied to technological change. This is Kuhn's idea of a scientific matrix (usually referred to as a paradigm).[16] The scientific matrix of a particular field consists of the people and institutions performing research in that particular field, the basic assumptions and underlying values of their theories, ontological models of the phenomena under study and preferred heuristic models, the particular theories and methods, the scientific instruments and laboratories, the data, a (usually) large number of unsolved puzzles, and one or a few texts that in a nutshell constitute the field of investigation by defining problems and showing preferred methods of solving them (this is the exemplar or the paradigm in a narrow sense). All these social, cognitive and material elements are normally woven together in a strong and coherent web: any anomalous element tends to be disregarded or reshaped to fit the overall context, and all new elements coherent with the rest are easily adopted while reinforcing the web. Only when the anomalies become more visible than the normal puzzles is a scientific revolution or change of paradigm possible. In many cases, the old and the new paradigm (the two are really incompatible at some level) live side by side and the number of scientists in each matrix decides which one will carry the day. Hence the new net has to be woven from scratch, although some of the work consists of reinterpreting data contained in the old web.

In using this framework as an inspiration for understanding technological research, two amendments seem appropriate. The first one concerns the exemplar, which is more likely to be a technological artefact (and its context of application) than a particular influential text.

The idea of an exemplar as a prime carrier of the entire matrix is particularly interesting. Research has shown that science and engineering students learn very effectively by working with an exemplar.[17] This is not a question of adding practical flesh to a theoretical skeleton. Rather, working with exemplary problems in science means to learn how to constitute problems so as to make them tractable under the appropriate theories. In other words, the internalisation of an exemplar is the same as adopting a particular outlook. The students learn how to make a situation look like one that is covered by the paradigm.

Similarly, engineering students working with exemplary artefacts learn what counts as a problem and what counts as a good solution. In both cases, the underlying values are transmitted without a need to formulate them. Successful internalisation means adoption of values and norms that – to some extent – only can be disclosed when they are infringed and subsequently attended to, as can happen in a design situation where a non-conformist or newcomer to the field has a strong voice.

The second amendment is to acknowledge the greater and more direct influence of external (i.e. from outside the profession) factors on technological design than on scientific discourse. Even though science is becoming increasingly applied, research in production technology is explicitly directed toward the perceived needs of the future customers, be they cost reduction or increased management control or any other need. This is to say that the validation criteria of technological research lie in the market (or its substitute, e.g. a continued user–producer relationship), whereas a scientific matrix contains its own validation criteria to a much higher degree.

As a corollary, opportunity may play an important role in making new technological paths. Economic theory largely acknowledges two driving forces of technological development: market pull (i.e. formulated needs of firms that buy and apply production technology) and technology push (i.e. technical opportunities seen by technologists with no user-defined needs yet attached to them). Hence, while scientific revolutions relate strongly to the focus on anomalies in the scientific matrix, technological revolutions may rely on external anomalies (i.e. the available technology is no longer suited to solving manufacturing or business problems), or they may derive from opportunities seen by technologists before a market need is developed.

What, then, would be key elements in transforming a matrix or replacing it by another; what are the conditions for redirecting technology? These would certainly include in the first place a combination of external anomalies (or dissatisfaction with the present technology) and new technological opportunities as seen by technologists. Secondly, with its key importance, an exemplary artefact that embodies the alternative set of values would be required. Thirdly, new theories, methodologies, instruments and the like would have to be developed around the new exemplar; but this would be a sign of its diffusion rather than a prerequisite. Fourthly, a social structure (perhaps a network) which accepted the basic tenets of the new paradigm would have to be built for its further development and acceptance.

If this train of thoughts and extensive reliance on analogies with science do in any way capture the focusing mechanisms of technological development, then the project has scored well.

As for the existence of, and focus on, anomalies, that goes without saying. In the academic community it has been the debates about skill and control, in the business community it has been the concern with dysfunctions of Taylorism, and in technological development, voices from both these external sources have reached in. Rosenbrock's own approach of considering Tayloristic technology as less than economical is, however, unusual among technologists. Nevertheless, the project shows that anomalies of the old paradigm are the object of attention.

The opportunity is also there, with software technology being almost embarrassingly flexible (to paraphrase Rosenbrock). I should add, though, that the flexibility ends after the software has been written; once finalised, it is as hard to change as cast iron, only even more brittle. After a few years of maintenance, recasting is the only alternative.

Furthermore, the project has hit an important target by aiming to produce an exemplary FMS for others to use as a model. Given the strong learning effect by copying and the cumulative character of technological development, a working demonstration is probably the most important single contribution at this stage. The other planned main product, the methodology for merging social science with engineering design, would be a secondary result according to the hypothesis; I shall return to this point below.

Nevertheless I feel uncomfortable about concentrating the efforts on developing an exemplar. I have three reasons for that, all of which are related to the immense capability of the technological community to reinterpret new elements into something old and familiar and consistent with the old paradigm. This is probably even more true with technological artefacts than with scientific facts.

While any particular piece of machinery carries with it strong suggestions as to its appropriate utilisation in coherence with the values embedded in it, it is usually not impossible to use it in other ways. For instance, a German study shows that conventional CNC work can be organised to leave programming with the operator, yet this is done in only 14% of the cases.[18] Rasmussen points to four determinants of the organisational choice in each firm: the product, the firm's position in the product market, the local labour market, and company tradition.[19] Obviously many of these factors are outside the control of the project (or any makers of technology). As for company traditions and responses to changes in the markets, however, these may be seen to be in strong (or sometimes weaker) coherence with the values underlying the technological paradigm, adding to the possibility of assimilation.

The Steering Committee also decided that, even had it been possible, they did not want to constrain the solution too much because they might end up with the wrong constraints. While this decision in my estimation was both wise and necessary, it did not make assimilation of individual bits and pieces of the software any harder. Technical artefacts tend to be almost infinitely dividable and at least there is always one part of them that can be extracted and put to other uses. In this particular case, Rosenbrock has already mentioned that manual data input systems are now commercially available that were developed from a set of values other than those of the present project. Both varieties carry the capability of being applied for constraining workers or putting them in control – although perhaps with different degrees of ease. Very little can stop the UMIST system from being applied as a temporary measure while waiting for NC programming to become an automatic CAD spin-off – except perhaps that extensive use might leave the companies with a workforce that was reluctant to let go of the newly-earned control.

Also in this case, the software support for the calculation of cutting data can easily be redone so as to remove the need for those skills in the machine operator and the office programmer. I believe that was the default assumption behind the original core development of that particular module before it was taken into the project. This is an example of how portions of an artefact can be removed from their context and be turned into their opposites, intentionally speaking.

My third concern is not with individual techniques in the software but with its entire constellation. The danger of assimilation rather than accommodation is greater when the new implement as a whole covers the same function as a previously existing solution and has fairly well specified interfaces to other functions. For example, one FMS concept can easily replace another FMS concept, or one programming language, once it replaces another, is easily adopted without further reverberations. This contrasts with the introduction of solutions that cut right across previously established functions. Consider the case of integrated CAD/CAM systems. Both CAD and CAM systems have fitted into the functional organisation of most firms, although not always in an unproblematic way. An effective integration between the two would, however, in many cases require firms to consider the design and making of one product as one function, and the design and making of another product as another function, and not in terms of designing all products as one function and making them as another. Hence the promised benefits of CAD/CAM integration are not achieved because CAD and CAM are assimilated into the existing organisational (and cognitive) structure of the firm.[20]

A main point here I think is the compatibility of a mature technology with the organisational setting in which it works. A technical solution that fits into that organisational pattern very easily reinforces the value structure (by confirming it rather than questioning it) and the organisational structure (by investing more in technology that presumes it). If an alternative technical artefact is to carry a widely different set of ideas, it might be better off if it does not slot easily into the existing cognitive and technical and organisational structure but rather forces some very conscious attention to all of these for its application. Such forced attention, I think, is not necessary for the introduction of an FMS or a CNC programming language.

To summarise, I believe the project has hit a crucial target in developing an exemplary solution that encompasses values more pertinent to what I shall call skill-based automation. Its efficacy in redirecting technology depends very much, however, on factors external to the project, to which I shall turn my attention in discussing possible future developments.

Future Developments

In this concluding section it might be tempting to say something about whether non-subordinating machinery or human-centred systems are more likely to form a paradigmatic core for the redirection of technology. I shall withstand that temptation, however. These concepts are still in their infancy and even if the continuation project under the ESPRIT project carries the latter term in its title while the present project title favoured the former, it is difficult to say how the two would have differed from each other had they gone in separate directions.

Rather, I shall give a glimpse of some other building blocks in the redirection effort, and specifically take note of potential allies and counter-forces. As noted in recent sociological investigations of scientific and technological development, the successful innovator is often the Machiavellian character who can negotiate equally well with the hardware in the laboratory and with the social structure that is expected to implement the new technology, hence effectively erasing the distinction between the social and the technological.[21]

The project itself has produced more building blocks than just the technological artefact. An important by-product was the rethinking of the value foundation of scientific knowledge, as documented here and elsewhere. In fact, without this more philosophical backing, the product of the project might be difficult to distinguish in kind from some of the commercially available systems which grant control to the operator as a temporary expedient.

When it comes to methodology, the project failed to produce a formal procedure for merging technological and social science knowledge for design. In this area, projects such as Utopia seem to have tapped the potential better, at least when it comes to developing a series of tool-like aids for facilitating worker involvement in design.

I think the achievements of the present project in the field of methodology have been understated, though. I can certainly read two messages. (a) The design procedure and the design criteria have to be understood jointly, not as two separate things. This message is corroborated by other findings[22] and runs contrary to traditional knowledge about design methodology. Hence a formal procedure is not to be expected that will work properly under varying conditions. This is of course true for traditional engineering design as well. (b) The control of the design should lie with someone who is more interested in redirection than in doing a good technical job. Every time a social science suggestion was not very well received with the engineering side, the burden of proof lay with the social scientist, whereas it was legitimate for the engineers to rely on their hunches as to what would work (technically or economically). To put it differently, the design engineers should be accountable to the redirector, whether the latter is an engineer or a social scientist. As far as I am able to tell from very brief insights, while perhaps not unanimously agreed, both these messages have become integral to the theory of action of some project participants.

What other allies can be enlisted in the redirection effort? Murphy describes the ESPRIT project elsewhere in this book (see Chapter 11), so I will not comment further on that. Cooley has mentioned some other research sites that are cooperating towards a common aim, enlisting each other (see Chapter 10). It should also be noted that the project has attracted unusual amounts of interest among social scientists interested in the possibilities of a changing technology. The potential of collaboration within this network has probably been underutilised.

Very few of these contacts have established independent projects that would actually create new technology; everybody seems to have waited to see how the present project fared, and what they have heard about for the last couple of years have largely been the difficulties involved in the enterprise. This is bad, because it probably takes a large number of independent but networked efforts to come up with sufficient momentum.

When we look to technological researchers, the interest has been far less, which is as expected from the paradigm point of view, although the support from ESPRIT was a major and important encouragement. Acceptance in engineering research circles is absolutely necessary, however, for survival. Senior automation design engineers in large firms may be a suitable target.[23]

Equally important is access to the institutions of engineering education. Appropriate influence on engineering curricula would mean that splitting (as defined by Klein, see Chapter 6) had been overcome at the social level. The concept of splitting is important and also quite useful because it combines a psychological and a social phenomenon. On the psychological side, people tend to organise their knowledge in personal constructs. This largely involves deciding on the main dichotomies in the world as one sees them and then clustering all one knows (rightly or wrongly) about, say, social scientists on one side and engineers on the other, constructing stereotypes that serve as interpretive schemes for incorporating new knowledge.[24] On the social side, similar things happen in the institutionalisation of different bodies of knowledge (which includes knowledge about other bodies of knowledge), the development of educational systems for maintenance of those differences, and so on.[25] While the project tried to overcome splitting within their own group by applying several forms of what Klein (Chapter 6) calls task 2 procedures, this psychological approach clearly is insufficient for addressing the social aspect, where appropriate methods are much harder to come by.

Unfortunately, experience from the movement of Science, Technology and Society programmes in the United States tells us that even institutionalised de-splitting may not be enough. A fair number of these educational programmes exist at many renowned universities. In my opinion many of them belong to one of two clusters.[26] One cluster consists of humanists and engineers-turned-historians studying the impact of technology on society. In this cluster scholars have come together from a previously split position to a fairly unified programme of research. The de-splitting is equally well achieved in the other cluster, which centres around technologists enlisting the services of social scientists in a fairly technocratic manner. We then have two substantially different outcomes of the de-splitting process, neither of which is what the present project was looking for.

As I have said before, though, the real test is in the acceptance of the technology by industry. In this respect, the problem of finding testbed sites tends to shift the control of the implementation from the researchers to the immediate concerns of industry. While this of course ensures solution of real world problems, it does not ensure that these problems are defined and attacked in a way which is coherent with the alternative paradigm. The reason for not finding many testbed sites is of course the inherent risk of trying out an unproved design. One solution might be to increase system reliability much further before offering it to industry, by which time there will be a larger potential group of interested parties. In this way, the purposive context of the laboratory (and not just its testing context) would have been extended to the arena of usage. The project then could choose a favourable firm, or a regional group of firms, or a branch of industry, or – if it was not for the national funding – even a country, if one thinks back to the studies I have quoted on cultural differences in technology development. A selection criterion might be that the traditions and commitments of the firms were supportive of a craft-based ethos and an anti-Tayloristic work organisation. Once again, this would not be a proving and debugging of the system but rather an attempt at establishing preferred patterns of development to be associated with the technology. I think this takes a much more massive effort than what can be achieved in a single testing of an early version in an environment that typically will tend to reduce the impact of the innovation on other routines in the firm, but it may be the best way to a more rapid diffusion. National technology diffusion programmes might be potential targets for achieving this.

Finally a note on counterforces. There is of course the whole problem of institutional inertia both within research, education and business, as outlined in the discussion of paradigm shifts. One premise underlying the whole project, however, has been that it is the value system of science rather than any particular political or economic system which is to blame for the present state of affairs. Hence it is assumed that an alternative technology has the capability of being accepted within the present industrial and economic structure.

Ullrich has published an intriguing analysis of the concurrent development of scientific technology and industrial capitalism which presents a less optimistic conclusion.[27] The main message is that capitalism and science, as they have developed historically and concurrently, have a structural affinity towards each other, and that this has led to a symbiotic relationship between them that has allowed for rapid development of both. As one instance of this affinity, the similarity between the scientific experiment and the capitalist production process has been noted: both should be conducted with full external control, and the internal processes are thought to be indifferent towards the external purpose. Needless to say, this short summary cannot give justice to the whole argument. If the analysis holds, however, then we might expect industry to be as opposed to changing the value system of science as science itself, because of cross-vested interests. This would definitely make the whole task much more difficult. We might need a strategy for dealing with any signs of such alliances once the non-subordinating or human-centred technology starts to gain acceptance.

References and Notes

1. In June 1982, I arrived in UMIST for half a year's stay with a fellowship from the Royal Norwegian Council for Scientific and Technical Research. My own aim was to study the influence of engineering design knowledge and practice on the qualities of the resulting production technology – in particular qualities that potentially affect skill requirements and control in the working situation. Methodologically it would be favourable to conduct such a study in a context where precisely that influence was explicitly discussed. Professor Rosenbrock generously agreed to have me in the position of a participant observer in his project, which was one of the two projects in Europe that to my knowledge could provide such a setting at that time.
2. Frederick Winslow Taylor, On the art of cutting metals, 1906 (American Society of Mechanical Engineers).
3. A great publicly funded research effort to determine optimal cutting data in four large Norwegian firms some years ago was only partially successful; one problem was that these data turned out to be specific to local context.
4. Michael Polanyi, Personal knowledge, 1958 (Routledge and Kegan Paul). Richard R. Nelson and Sidney G. Winter, An evolutionary theory of economic change, 1982 (Belknap Press of Harvard University Press).
5. Jean Piaget, The principles of genetic epistemology, 1972 (Basic Books).
6. Marvin Lee Minsky, The society of mind, 1986 (Simon and Schuster).
7. Dr L. Bainbridge, personal communication.
8. Chris Argyris and Donald A. Schon, Organizational learning: A theory of action perspective, 1978 (Addison-Wesley).

9. See for example Harry M. Collins, Changing order, 1985 (Sage), on the difficulty of replicating experiments without having really taken part in the original ones. Polanyi (see ref. 4) also stresses heavily the personal character of scientists' knowledge.

10. Pelle Ehn, personal communication.

11. Håkon Finne, A case study on the development of a flexible assembly system, in K. Rathmill and P. MacConaill (editors), Computer integrated manufacturing, 1987 (IFS/Springer-Verlag).

12. See i.a. Peter Mambrey and Barbara Schmidt-Belz, Systems designers and users in a participative design process, in U. Briefs, C. Ciborra and L. Schneider (editors), Systems design for, with, and by the users, 1983 (North-Holland), for similar findings.

13. Herbert Kubicek, User participation in systems design, in Briefs et al. (see ref. 12), summarises some of the context variables that seem more important than the actual systems development tools.

14. Philip G. Herbst, Community conference design, Human futures, summer 1980.

15. Håkon Finne, Henning Neerland and Tore Nilssen, Bedre montasjejobber? 1984 (Institute of Social Research in Industry, Trondheim).

16. Thomas S. Kuhn, The structure of scientific revolutions, second edition, 1970 (University of Chicago Press).

17. Arne Jakobsen, personal communication.

18. Helmut Rempp, The economic and social effects of the introduction of CNC machine tools and flexible manufacturing systems, in L. Bekemans (editor), European employment and technological change, 1982 (European Commission).

19. Bente Rasmussen, Fagarbeid og ny teknologi, 1984 (Institute of Social Research in Industry, Trondheim).

20. Håkon Finne, Organizational alternatives in the integration of CAD/CAM, in Michael Rader, Bernd Wingert and Ulrich Riehm (editors), Social science research on CAD/CAM, 1988 (Physica-Verlag).

21. Bruno Latour, Science in action, 1987 (Open University Press).

22. Frits Prakke (editor), John Bessant, Håkon Finne and Hartmut Hirsch-Kreinsen, Human factors in systems design: methodology and cases in factory automation, 1987 (TNO, Apeldoorn).

23. Upon the initiative of the international production engineering organisation CIRP, a working group is putting forward a research proposal to the ESPRIT programme about design methodologies and criteria for human factors in computer integration, and they aim specifically at senior designers as a target group for their research results.

24. Donald Bannister and Fay Fransella, Inquiring man: the theory of personal constructs, 1971 (Penguin).

25. Peter L. Berger and Thomas Luckmann, The social construction of reality, 1966 (Doubleday).

26. When gathered at their 1987 conference, most delegates seemed to be more concerned with maintaining their programmes in face of university cutbacks than with integrating disciplines; redirection of technology as a goal was not mentioned once.

27. Otto Ullrich, Technik und Herrschaft, 1979 (Suhrkamp).

Chapter 8

A Works Director's View

Allan Chatterton

Introduction

Over the past 16 years I have been involved at senior executive level in the total manufacturing operation at a medium-sized electrical engineering company based in the suburban district of Manchester. The company's overall performance over this period has been well above that achieved on a national scale in the manufacturing industry. In looking back over this successful period the one element within the company that has remained stable, predictable, committed and flexible is the people element. The management team without exception has attached both a human- and task-centred approach to all decision-making with social and economic effectiveness.

In the early 1970s most companies in the UK produced to a design that was well established, and manufactured by conventional production methods and machines. The rise in oil prices in 1972 followed by expansionary measures in the UK in 1973 by Prime Minister Heath and similar measures by leaders of the industrialised world, generated a period of inflationary growth.

Whilst companies in the UK were striving to overcome the 'Us and Them' syndrome in our factories, exacerbated by increasing wage demands to match inflation, management ignored the threat of the competitive nature of the newly industrialised nations. There was a non-realisation that product innovation was stagnant and products generally in the UK were moving very quickly into obsolescence. By 1978/1979 the various changes in government in the Western world were beginning to have an effect on the direction of the economy. Attempts to squeeze out inflation caused commodity prices world wide to fall, and only the fittest in manufacturing and other industries survived as the recession gathered momentum. Industries suddenly realised that their product portfolio, manufacturing methods and control were in need of rapid change and restructure. Cash flow, reduced inventory and work in progress, together with short batch work and improved customer response, for those companies who had a marketable commodity, became the order of the day.

This precipitated a surge forward into the minefield of production methods and manufacture based on new technology machines which promised to be the immediate answer to arresting the decline.

Throughout the 1970s we as a company were able to perform successfully within these cyclic changes but, in the late 70s and early 80s we were thrust into a price war with the competition in a buyers' market. The message to industry at this time from the politicians, the press, the city and the leading economists was that all the right things industry was doing in the 1970s would now be the wrong things in the 80s and 90s and the only way forward would be the speedy introduction of computer-aided manufacture in the forms of FMS, CNC machines supported by DNC computer networks, etc.

Evolution of New Technology and Human Skills

In my experience the introduction of new methods with the cooperation of the labour force has often proved to be successful. As a company we have ignored the pressure to automate as rapidly as possible and concentrated on finding ways to develop and install the best technology that improves life on the job and provides efficient production.

New technology rather than contributing to the control of the production process can be beyond the capabilities of most companies to control. Attention should be paid to the technical development of the new facilities, and adjustments in the organisation, needed to accommodate a social technology approach to their usage. Rather than the customary procedure of fragmenting jobs and developing specialist people in the planning department, the operator's skills should be fully utilised. Perhaps more important, his talents should be extended to additional fields as an alternative to selecting a skilled operator to carry out prescribed tasks, as controlled by the methods department, which will lead to stagnation of machine skills in the future.

Evolution of technology should allow existing human skills and abilities to remain relevant but should also allow these skills and abilities to evolve in a way that matches the evolution of technology. Robots, CNC and FMS etc. are not replacements for people – they should be viewed as an element in a system in which human abilities and machines are complementary. The systems need to be designed so that they satisfy the technical, economical and human requirements. They should not be dependent upon a few key experts whose absence might temporarily disable the systems. Management should pay less attention to cost and output ratios per man-hour employed and focus on profitability from ease of product change and market response, leading to reduced inventory and shorter lead times.

Management do not want to experience the immense difficulty in coping with cognitive and motivational problems that emerge very early from the acquisition and use of new technology. Disorientation can be induced in individuals by subjecting them to too much change in too short a time. Taking an individual out of his own culture and forcing him into an environment sharply different from his own can have serious consequences on a successfully fine-tuned structure within any company. There must be a balance between the pace of change and the limited pace of individual response. Our choice of technology in the UK and the approach to implementation will decisively shape our work style and sub-culture of the future. The previous fragmentation of worker tasks should be reversed. The increasing educational level of workers requires us to look carefully at job enrichment and job education under which workers can be delegated the responsibility to control their own working conditions and environment. If employees' knowledge is allowed to be depleted as a result of knowledge being manipulated and standardised, stagnation of social development will ensue together with a reduction in the ability to invent and develop new products and methods. It is vital that machines based on micro-processor technology constitute an aid to processing leaving the scope for initiative and creativity unreduced.

If a period of technical change is not looked upon as the right time for learning more about employee involvement and co-determination of systems and product designs etc., introducing new technology in a company will lead to unplanned side effects which will hamper effective use of any machine based on new technology.

To achieve better results from investment into new technology then we must endeavour to fulfil the following:

1. Social criteria as outlined above must be acknowledged. It must be remembered that without the proper attitude on all sides of production a satisfying result cannot be negotiated.
2. Early involvement by the users of the system at the time of conception and development is most essential.
3. An agreed re-training programme must be developed immediately after choice of new technology is made.
4. Management at all levels must be kept informed of company-wide developments and strategies applied to new technology. Without their support a stable situation will never be established.

The main purpose of introducing new technology is to provide a flexible and rapid response to changes in market models and production bottlenecks. Higher efficiency must result if operators are allowed to act quickly, communicate change and correct errors in computer control. The number of eyes and ears sensitive to the detection of machining problems will be expanded and the competence available to deal with a broader range of faults will be extended. A higher collection of expertise will allow easier changes to the parts, components and models that design improvements or market changes might dictate.

The Operator, His Machine, and a Human-centred Approach to Technology

For programming and operating machine tools tacit knowledge is needed which cannot be substituted for by technology. Automated assistance for small batch work should be designed around this available knowledge to utilise it to its maximum.

Before the arrival of computer-based automation, the machinists' skills provided that missing link between a blueprint and a finished part. Using his innate skills he would be careful to make adjustments to the running of the machining conditions and create useful fixtures and accessories. There are many compelling technical reasons for programming of machines to be done *off* the shop floor, but devising long complex programs will require many intricate calculations, which will take several man-hours. It is more effective for operators to develop programming themselves with computer assistance. They are well situated for debugging and improving programs to enhance machine efficiency. The higher the operator's skills the more you get out of the machine.

It has to be assumed that the operators responsible for automated machines have the ability to read drawings and discuss other geometric information. The mental picture of a machining sequence can then be formed and, when combined with tacit knowledge, be transmitted into an operational sequence to form the basis of an input into a computer interface.

The interpretation of drawings into machine operations, deciding how set-ups will be carried out, determining methods of holding work, what tools to use and in what sequence, what speeds, feeds etc., should be open to the operator, allowing any computer control function and its programs to perform concurrently but always under the guidance of the operator in charge.

The UMIST Project

The UMIST FMS project in utilising human criteria does not impose or pre-empt how the users of the system will operate. Rather it identifies known areas of unpredictability and defines the parameters of uncertainty within the system, thus allowing the operator to change his perception of the problem or task in an endeavour to reach the best situation without losing software support. The system provides program generation by the operator using conversational interaction, colour graphics etc., allowing the operator to be released from much of the physical work and giving more time for him to study the development and improvement of the system. This provides optimisation and extends operator skill in many additional ways.

The operator is given the opportunity of increased choice, i.e. becoming involved in tasks such as scheduling, tool changeover reductions etc. This increases his command of the local manufacturing situation, which can only improve overall efficiency by the extension of the areas where the operator can exercise his engineering judgement.

The project at UMIST in designing and developing a FMS utilising human–machine design principles has the following desirable criteria:

1. It does not subordinate workers themselves but instead acts as an aid to their skills and ability to be more productive
2. It is technologically satisfactory in performance and reliability
3. It is economically satisfactory and attractive for use in medium and small engineering environments

Conclusion

Looking at new solutions and applying a fresh approach opens the way to improved working conditions, a reduction in excessive administration and improved cost effectiveness. By applying a human-centred approach success can be very convincing, giving considerable economic advantages combined with job enrichment, and managers can begin to assume the role of advisers and instructors. Workers can become more responsible for ensuring production flow rather than being limited to caring for the simple task of speed in component machining. If they reject the capability of men and women to react to the unforeseen, and to overcome difficulties through a sense of purpose, systems can quickly become rigid, inflexible and insensitive. The quality of work suffers, levels of efficiency suffer and employee motivation suffers.

Total dedication to using the company's wealth of employee intelligence to harness technology and develop those amazing inner resources is the true responsibility of management.

Chapter 9

The Coordinator's View

Harold Palmer

Introduction

I first came into contact with Howard Rosenbrock's FMS project when he applied for research funds to the SERC/ESRC Joint Committee for whom I acted as coordinator. Subsequently, I attended, informally, two meetings of his Steering Committee: I was then asked to join his Committee and attended most meetings, so that I was able to follow the project from start to finish.

I was attracted to the research by the triple aims of, first, establishing the principles in an academic environment for a new technology, secondly, bringing social and physical scientists/engineers together in experimental research and, lastly, writing an interactive 'human-centred' computer program for craftsmen to use.

I am an industrial chemist, with a background of managing the introduction of new technology where the people and their ways of working are at least as important as the equipment. I am convinced that involvement of people at all stages of an innovation is vital. Without this involvement, innovations are rarely successful, businesses do not develop and people become alienated.

I believe that the project was successful, and I would like to amplify this conclusion by moving from the general achievements, through the conduct of the project, to the problems of engineers and social scientists experimenting together to develop technology.

General Achievements

Sufficiently well defined new principles for a turning cell have been established for makers of lathes to be interested in further technological and commercial development and for ESPRIT funding to be awarded. This support to use the new technological principles has been won by the combination of engineering, social science and computer science in the project, and by the way in which the researchers have sought to achieve a worthwhile result together, despite their differences. The researchers consistently sought to achieve a result, through experimentation, which met not only the aims of the project, but also their own professional and personal needs. Alone, a team of researchers from a single discipline is unlikely to have achieved the results within the constraints of time and of resources. Possibly, systems analysts with a basic knowledge of engineering and sympathy with production people could have written a suitable program, but this program would probably not have been acceptable to craftsmen and few new cross-disciplinary principles would have been highlighted.

This establishment of technological principles has gone hand in hand with a search for ways for social scientists to work with physical scientists. The efforts have been good, but the results are limited and, with the break-up of the research team are unlikely to be used elsewhere either in research or in development, except by the people actually involved in the project. This limited use comes in part from the nature of science and scientists in general, who seek great certainty in what they do, and partly from the state of social science, which has not yet been able to establish principles and knowledge to compare with those of the physical sciences: there is also the greater variety of human characteristics and situations.

The project was well organised within the resources and the academic aims of UMIST: the people's skills were well used within the research team, on the Steering Committee and outside UMIST where people had relevant skills or might use the final technology. Finally, the researchers gained in skills and abilities through the project especially, naturally, in cross-disciplinary working and in relating to people outside the project.

The Conduct of the Project

Compared with similar projects within manufacturing industry, there were a number of differences in the way in which the project was conducted. Perhaps the most important, and the one which could have only been overcome by having a close working relationship, as in a CASE award, with an industrial company, is the fitting of the research into a business strategy. The emphasis was on writing the software for the turning operation. A nod was made towards costs by assuming that the cost of computer hardware is falling rapidly and so there would be cost advantages. No account was taken of the effects on the costs of either the company that might use the results of the research or that company's customers. In industry, as soon as the software was developed, work would have started on things such as job design, work organisation, production planning and marketing; the absence of consideration of pay and of costs made for unreality. Presumably, just from developing the system, there would have been cost disadvantages to the company using the new production and marketing system. However, there would then be benefits to that company of more effective people as well as benefits to their customers, such as better quality and more consistent delivery. Any good innovation has to be developable beyond what it is when first introduced, and the scope for development has to be seen through people in production and sales within the company as well as within customers. The design of software achieved by the UMIST team is clearly sufficiently robust and developable for the ESPRIT programme to start speedily. Such robustness and developability are the characteristics of a good design. Now, three years on, with the completion of the first stage of the ESPRIT project, the merits of the UMIST design are well proven and cost as well as business advantages have been confirmed.

The abilities as well as the time of the researchers were underused because the experimental facilities of UMIST were limited. Not only were the researchers kept waiting because equipment was not available; new ideas could not be tried as new results were obtained, because experimental rigs were not provided. The project needed support from much more experimental mechanical engineering: for instance the development of new mechanical handling to support the craftsmen could have gone on alongside the social science work and the writing of the computer program.

Lastly, the researchers and, in a different way, the Steering Committee were over concerned with conceptual dogmas which have been formulated by academics, clerics and politicians. As an industrialist, I do not decry these ideas, but, in the project, there was a tendency to discuss rather than to experiment and to give greater weight to the published ideas of other people. Instead, there should have been more emphasis on the team's knowledge and expertise as well as on the needs, aspirations and aims of craftsmen. I have seen the same tendency to emphasise dogma at the expense of experimentation and need within the central laboratories of large industrial companies. With such central laboratories, the rest of the company has to work hard and sympathetically with the laboratory management to relate the research to the business.

Cross-disciplinary Work by Engineers and Social Scientists

The successful collaboration between people from different scientific disciplines in this project has illustrated the problems of finding engineers and social scientists who can experiment together to develop new technology. The engineer wants a specification to follow, while the social scientist wants to investigate a human situation in depth without influencing it. In industry, where a social scientist may help with organisation development or with ergonomics, there is usually neither the time nor the ideal experimental conditions to establish new principles in social science. A social scientist is also often reluctant to translate a successful experience to a different technological situation, whereas an engineer has to produce a new design or system quickly to meet a series of new needs. Very special outward-looking people are needed from the different disciplines.

In industry, enough new experimental results are generally obtained for a design team to go ahead with confidence. This semi-empirical approach needs strengthening by much more cross-disciplinary experimental research on the basic principles.

The Importance of Management and of Technical Process Development in Human-centred Working

To help to put my conclusions about the UMIST project into context, I would like to make a few general remarks about management, process development and human-centred working.

Good management plays a key part in maintaining and improving human-centred working. Change is continuous within a working team: even with a constant demand for the output as well as with the same equipment and ways of working, teams as well as individuals learn how to perform better, and so their effectiveness improves. This learning eventually means that the balance of skills that are needed changes within the team and so the team has to reorganise. With a rising demand either for greater output or for improved output, this reorganisation can take place gradually and naturally. Sometimes the reorganisation needed is so fundamental that completely new skills and ways of working are needed. Here a perceptive and skilful management has to carry through the changes. Often, the people within a team can see the need for change before senior management, but rarely can they be articulate or specific about what is needed. Where working relations are already good, everyone can get together to evolve the changes. Where the relations are not good, management still has to take the initiative and bring about the necessary changes. These changes frequently alter the balance of power between individual people as well as between different parts of an organisation.

For instance, within a small engineering workshop the craftsman is often dominant and so he decides priorities for the other people in the workshop as well as for the sales people and even for the owners. As the workshop enlarges, a production control unit may be needed and this unit begins to dominate. Frequently, the production controllers are ex-craftsmen who have the skills in numeracy and in self-expression, which are needed. They are now 'white-collar' workers and also called controllers! They will be among the most reluctant to accept a subsequent change in their role and status.

This striving for dominance by individual people or functional groups can be carried on throughout an organisation. Just as the maker can give way to the production controller as the demands on the organisation change, so the production controller can give way to the sales controller, who in turn gives way to the marketeer. Such changes in dominance can take place with no change in basic technology, as demand grows and the product matures. Individuals can feel lost while these strivings go on in a changing way around them. Management too has to learn how to stand back periodically to perceive and to carry through the changes that are needed.

Within these organisational changes, a craftsman often has to change his role more often than other workers, including management. For instance, when a product is new, only he has the skills to make first the prototype and then the early production. As the product and the mode of making become better defined, his skills may not be needed and less skilled people can take over the making. Within a good organisation, the craftsman will have moved on to another role and neither he nor his skills will have been abused. Too often, of course, neither management nor he perceives the need for change, and frustrations develop.

So far in this discussion of changes with time in dominance within manufacturing organisations, I have not mentioned the role of technical development. Within the normal ongoing learning processes, small technical changes will be used which will have been developed either in response to a perceived need or as a result of new technical information and better technical understanding. Ideally, the production unit should be a technological laboratory in an unobtrusive way, where better understanding is obtained, not only about the technical processes involved but also about human skills, needs, aspirations and satisfactions. Such an ideal is always feasible without disturbing production, and, indeed, production must not be hindered, but often production benefits from such attention as a laboratory. Mutual trust between everyone is needed: this trust does not mean that everything is happiness and light; uncertainties, misunderstandings and conflicts inevitably arise, but trust and a common aim of improvement mean that eventually everyone benefits.

The technical changes which are part of the ongoing learning process are often minor, such as new pallets to hold work in progress, a new sequence of assembly or new connectors to eliminate a soldering stage. Even so, these minor technical changes can alter the role of a skilled person such as a wirer, miller or welder. Equally, skilled persons can be frustrated if the right minor enabling technology is not available for them to do a decent job. Even more drastic changes for production people can come from the technical drive to improve the quality of existing products. As processes and products become more reliable, so the role of the operators changes, in turn, from making, to intervention based on their skill and training, to minding and, then, to observing and recording. Different sorts of people are needed for these different sorts of jobs; they must be organised differently and their jobs must be designed differently. Overnight, by improving the quality at one stage of a production process, the need for a skilled person at that stage can be eliminated, so that the mix of skills and hence the 'pecking order' in the production team changes.

These sorts of changes in people-centred working are the result of a striving for improved product quality: in that sense they are technically driven, but a perceptive working team will aim to give the people involved in the making of the more uniform products a meaningful job. The team will also ensure that the people whose skills have been supplanted are given meaningful work perhaps in a different production unit. Generally, craft, skilled and semi-skilled people are in short supply. However, sadly, they are often unwilling to change roles or organisations, and, also, management is content with a seeming status quo which is, in fact, changing irrevocably and disastrously.

The elimination or simplification of a stage in a production process is another sort of technical change which can alter either the skill or the status needed for a job: often too production stages are telescoped. The use of connectors instead of soldering has already been mentioned: adhesives have transformed many manufacturing processes by eliminating joining operations such as stitching, riveting, bolting etc. Too often, management has expected, say, either the old stitchers to apply adhesives without regard to their previously hard-won skills and pride, or school leavers to apply the adhesive to give the same neat job as the old stitchers. Unhappily too, the old stitchers have been reluctant, even with a perceptive management, to learn new skills and new satisfactions; the school leavers also have resisted being told how to do a simple straightforward operation.

So far, relatively small technical changes in manufacturing processes have been considered, but the same basic principle of the involvement of people applies equally for large changes as for small. The wholesale elimination or simplification of stages in a production process can involve massive technical development and investment of capital. Within the chemical industry, the progression from craft-based batch processes, through batch mass production to continuous integrated processes is typical. Often the resources needed for this progression have been so large that the management of these financial, technical and organisational resources has been the dominant concern of senior management. In this management of resources, senior management forgot the people who had to build and to operate the final continuous process plants. Senior management now knows that such large-scale process innovations are only successful when these builders and operators are kept in mind and involved throughout the innovation. In other words, people have to be kept in mind throughout a process innovation, and technical process innovations have a major bearing on the way in which people work together effectively and obtain satisfaction.

Within manufacturing industry, the interactions between peoples' hands and brains are vital to the effectiveness and satisfactions of individual people. These interactions also contribute to the overall effectiveness of an organisation and good tools help the hand–brain interactions. Since the UMIST project was concerned with craftsmen, I will confine my remarks to craftsmen, although I am reluctant to do so, because my training and experience in their work is very limited. I will, however, comment at some length as I think that the improvement of this kind of interaction is vital and was a prime design requirement for the work at UMIST.

To me, the mental attitude that a person has to a job is vital to the effectiveness and satisfaction of that person. A craftsman operating a lathe is not just working through his sight and his numerical setting of the lathe. He is taking in information from all his senses and feelings: he is listening to the sound coming from the work in the lathe, the vibrations he feels as he operates the controls and so on. He then integrates all of this information into a mental picture of what is going on and of how the work will turn out. At the same time, he is thinking backwards as well as forwards about how the work came out previously and of how he might improve the work in the future. He is also not just thinking about the current turning operation: he is thinking about things such as how the pieces come to him for turning, how the work leaves him and, at times, how the work is used by the final customer. One purpose of an apprenticeship is to develop this overall awareness and effectiveness through the integration of all his abilities in thinking about how to work more effectively.

Any equipment, procedure or training which will improve his performance in the whole of this sensing, analysing and synthesising is welcome to him. The improvement may come in unexpected and ill-defined ways, so that experimentation and an open mind are vital. Some people, outside manufacturing, view new equipment and systematic procedures as soul destroying. Sometimes within manufacturing they are seen as menacing. Properly used, they, like any good tool, can increase, in dramatic ways, people's effectiveness and satisfaction. However, certainly, if wrongly used, they can stultify and misuse people's abilities.

Generalising, we can say that the brain monitors and guides the activities of the worker's body and of the brain itself on the basis of old information which is held in an orderly way within the brain. The brain is constantly reviewing not only the information, but also the way in which it is ordered. As part of this ordering, the worker's brain contains models based on past information: these models include past empirical information about external events as well as the results of actions by the person in response to these events. Paradigms and concepts play a part too in the building of these models.

The brain continuously monitors and guides on the basis of these models. In turn, the brain is stimulated by new empirical information gained by the body's senses. If the senses are not monitored by the mind, they will become uncontrolled and confused. If the senses are overcome by the mind's need for exactness and neatness, the work will become arid and lifeless. This stimulation by new empirical information leads the brain to new models and a search for new information. Without such stimulation, the monitoring and guiding become increasingly stereotyped as well as inappropriate.

Everyone, of course, has these interactions between the mind and the senses and here new empirical information is vital, especially to people concerned with making rather than theorising. Often, too, the rethinking results from activities outside the job in hand.

One strength of the UMIST project was the new empirical information that was won to write the program and the way in which this program will give new empirical information to help the decisions and actions of the operator as well as the managers who use the program.

So far, in this general discussion of the interaction between management, process innovation and human-centred working, I have dealt largely with need-driven and relatively minor innovations such as improved product quality and better use of human skills. Occasionally, large process innovations are needed in unexpected ways to take advantage of a technical invention. The new process or product may need levels of precision not hitherto thought possible or there may be hazards which involve new ways of working. Transistors are an example of how greater precision calls for new ways of working and nuclear power of the way to control new hazards. Such radical changes in working procedures and skills properly managed give opportunities for new, more satisfying human-centred ways of working. The best practices in these two industries have lessons for everyone in the older manufacturing industries.

Finally, I would like to make some general comments about the role of computers in achieving more human-centred ways of working. My view simply is that there are many untold ways of increasing human-centred working within as well as outside manufacturing industry by the application of computers. I know of no application of a computer to improve a process where the benefits have not been greater than anyone dared to hope and where people have not welcomed the new computer-based processes. The reasons for such an optimistic generalisation are, first, that the use of the computer has forced people to think deeply as well as quantitatively, perhaps for the first time, about the basics of the process. Secondly, inevitably, bringing in the computer involves everyone in the production process. Lastly, minor enabling technical innovations are made, e.g. sensors to control the process, which improve the production.

With the wider availability of computers and greater general knowledge about computing, increasingly a computer expert is not needed and, often, it is better to lose some superficial efficiency, so that the non-experts learn and then become involved: the expert is there to advise and enable, not to dominate. The last thing needed is a new group of obscure experts to operate the new system. The computer system can make old functions such as production control obsolete, so that new organisational power structures evolve. Because so much information is available in a systematic form, new perceptions can form and new things can be achieved. Commonly, the quality of the output rises, sometimes, as discussed earlier, to the detriment of some individual people's skills, but generally to everyone's greater satisfaction. One benefit of the UMIST lathe program, for instance, was that toolroom craftsmen had less spoilt work in the machining of prototypes and so their satisfaction improved. This improvement was not anticipated and is typical of what is achieved when computer-based systems are introduced.

Another common benefit is that the level of work in progress falls so that stock holding, production planning and response to customers' needs are all improved. So far as individual workers are concerned, there is less waiting for other workers and there is less interference from people with ill-defined unaccountable functions, who hinder the actual production line. Everyone should be able to see more clearly their place in the activities of the organisation, none should have diminished roles and some should have wider jobs. For managers, one benefit is that they can get together with their people to manage their function, because, often for the first time, they have quantitative information instantly available in a form that everyone can understand. With a well-designed system, there is no need to review everything systematically: instead, exceptional happenings can be dealt with as they arise. A good manager, like a good craftsman, has interactions between hand and brain which can be enhanced by a good computer system to increase his effectiveness. Finally, the technological learning and the speedy response to customers' needs, which are so vital to success in production teams, can go on more speedily with the aid of the computer. Information about past performance as well as about potential performance is so comprehensively and clearly available that people can see how and where to improve their performance.

The Future

There is a role for Academe to establish the appropriate knowledge and skills with industry for more human-centred working systems. Interactive computer-based systems are needed in a variety of human situations not just in manufacture, and the technology is new, so that there is a chance for cross-disciplinary teams to work together not only in establishing the new technology but also in having this technology accepted by the users. Howard Rosenbrock has made a start, and there is often good isolated cross-disciplinary work, but much more systematic user-orientated research is needed to establish the principles for more human-centred working systems.

Final Comment

The brevity of this chapter reflects my industrialist's way of writing. We all worked freely with Howard Rosenbrock as individual people committed to a common purpose, and we all expressed ourselves naturally. I hope that readers as well as my co-authors will understand that my brevity underlies a deep commitment, not only to the way in which the cross-disciplinary research was carried out, but also to the multi-faceted outcome of the work. The basis was laid for a human-centred system of operating a lathe through a computer, and a significant start was made on finding ways of bringing, in an explicit way, human needs, aspirations and talents into the design of manufacturing equipment. I believe too that each of us was enriched personally and professionally by the experience.

Chapter 10

Human-centred Systems

Mike Cooley

Introduction

When human atoms are knit into an organisation in which they are used, not in their full right as responsible human beings, but as cogs and levers and rods, it matters little that the raw material is flesh and blood. What is used as an element in a machine is in fact an element in a machine . . . The hour is very late, and the choice of good and evil knocks at our door.[1]

The Early Stages

The past thirty years have seen the critique of science and technology shift from narrow fringe groups to still small but significant academic research projects, practical exemplary projects and, in the case of West Germany, a significant political movement linked to ecological questions. The early stages were characterised by the anti-technology culture of the 1960s, to be followed by the anti-science movement of the seventies. The movements were strong on criticism and weak on solutions. They tended also to concentrate on rejecting the products and values of industrial society, at least at the rhetorical level, concerning themselves with the impacts on consumers whilst ignoring the impacts on the producers. According to their leading activists, science should be viewed as evil, totalitarian and devoid of those attributes which might make it amenable to the 'Human Spirit'. This position is expressed clearly today by those such as the German writer and radical Bahro, who broadly asserts that the values of science and technology run counter to the human essence. In parallel, there were less vocal but potentially more significant critical developments at work. I will mention but three of these.

Technology, Science and Ideology

First, within the Marxist tradition, there was a growing dissatisfaction with the Bernalian analysis of twentieth century science.[2] In this analysis, science, although it was integral to capitalism, was ultimately in contradiction with it. Capitalism, it was felt, continuously frustrated the potential of science for human good. Thus the problems thrown up by the application of science and technology were perceived simply as capitalism's abuse of their potential. The contradictions between science and capitalism were viewed as the inability of capitalism to invest adequately, to plan for science and to provide a rational framework for its widespread application in the elimination of disease, poverty and toil. Within this view the forces of production, and in particular science and technology, were viewed as ideologically neutral, and it was considered that the development of these forces was inherently positive and progressive. It was held that the more these forces of production, technology, science, human skill, knowledge and abundant dead labour (fixed capital) had developed under capitalism, the easier it would be to effect the transition to a socialist society. Those who questioned this orthodoxy, both inside and outside the socialist societies, were held to be somewhat irrational and non-progressive, since science, at least in its own terms, is rational and progressive. Science had after all, so the argument ran, destroyed the human-centred model of the universe through the Galilean revolution, and made redundant through Darwin earlier ideas about humanity and the creation of life. Science viewed thus, appeared as critical knowledge, liberating humanity from the bondage of superstition; a superstition which, elaborated into a system of religion, had acted as a key ideological prop for the outgoing order. By the early seventies there was a growing questioning of this rather mechanistic interpretation of the Marxian thesis. There was a gradual realisation that science had embodied within it, many of the assumptions of the society which has given rise to it. This in turn resulted in a questioning of the neutrality of science as it is at present practised in society. Such questions extend far beyond that of a scientific use/abuse model to deeper considerations of the nature of the scientific process itself. Science, carried out within a particular social order reflects the norms, ideologies and culture of that order. Science therefore, ceases to be seen as autonomous, but instead, part of an interacting system of which internalised ideological assumptions help to determine the actual experimental designs and theories of scientists and technologists themselves.

Secondly, there was questioning in academic circles prompted largely by cybernetics, computer systems and a plethora of 'new technologies'. Norbert Wiener's *The human use of human beings* appeared in 1950.[3] The book appeared in various forms throughout the fifties and sixties, and during that period Wiener wrote:

I should like to dedicate this book to the protest against this inhuman use of human beings. It is easier to set in motion a galley or factory in which human beings are used to a minor part of their capacity only, rather than to create a world in which these human beings may fully develop. Those striving for power believe that a merchanised concept of human beings constitutes a simple way of realising their aspirations to power. I maintain that this easy way to power not only destroys all ethical values in human beings, but also the very slight aspiration for the continued existence of humanity.[4]

Ida Hoos produced her challenging paper 'When the computer takes over the office' in 1960. By 1965, Hubert Dreyfus was comparing artificial intelligence to mediaeval alchemy. His critique of AI was developed and culminated in the book *What computers can't do*, published in January 1972. Also in 1972, Joseph Weizenbaum produced his *On the impact of the computer on society* with the challenging subtitle *How does one insult a machine?* By 1976 these concerns had found fuller expression in his *Computer power and human reason*.

Of particular significance during that period was the gradual transformation of sociological, philosophical and political critiques into alternative systems design proposals. The American design methodologist Professor Lobell cautioned that design was a holistic system and it was necessary to provide space within such systems for those attributes which do not accord with discernible forms of logic systems.

It is precisely for that reason that the most powerful logics of the deep structures of the mind, which operate free of the limits of space, time and causality, have traditionally been responsible for the most creative work in all of the sciences and the arts.[5]

Proposals for a New Approach

In 1973, Bob Kling proposed a 'people-centred computer technology' and the IFIP conference 'Human Choice and Computers' in Vienna in 1974 considered several papers beginning to suggest alternatives to the given systems design. During the same period, Rosenbrock had been working on the use of computers in systems design, and his *Computer aided control systems design* appeared in 1974, to be followed in 1976 by his seminal paper 'The Future of Control'. The writer's own book *Computer aided design, its nature and implications* appeared in 1972 and was followed by a series of related papers and reports throughout the seventies.

The third development was that amongst industrial workers and the trade unions. The optimism of the sixties, expressed in the huge conference 'Qualität des Lebens'[6], was gradually transformed, in particular at factory level, into a questioning of what was made and how it was being made. Fertile discussions took place amongst the Fiat workers in Italy, those at LIP, in France, Algotsnord in Sweden[7] and perhaps the most renowned, the Lucas workers' Plan for Socially Useful Production.[8]

Whilst that Plan is probably best known for the range of alternative products described in it, of equal importance to the Lucas workers was the notion it embodied of systems which enhanced human skill and ability. This found expression in their proposals for telechiric devices which, as they put it at the time, were 'Manufacturing and Maintenance Systems which provide an audio-visual and tactile feed back, and which respond in real time to the skill and ingenuity of a worker, but do not objectivise that knowledge'.[8]

The significance of the UMIST project was that it represented, for me, a convergence of these three developments or strands. More particularly, it provided a framework in which a system could be designed which would demonstrate *in practice* the potential for human-centred technologies. The prospect of transforming the critiques and theories into practice excited the engineer in me and in any case, as Goethe put it in *Faust* 'All theory, dear friend, is grey, but the golden tree of actual life springs ever green.'

I did not regard the project as a 'one shot' dramatic activity. It would for me be one building block in a new structure. When an alternative is being developed which challenges the given orthodoxy, one is engaged in a long-term process. It is ultimately necessary to demonstrate that the new system is significantly better than the old, otherwise a change will not result. Such is the nature of paradigmatic shifts. The established methods never have to withstand the same destructive criticism and hostility as the new ones. The existing technology will have around it an infrastructure of manufacturing techniques, design philosophies, educational programmes, all of which are used consciously or unconsciously to support the existing forms. Innovations, however radical a departure they may be, will always be treated sympathetically and with considerable tolerance if they are seen to be swimming with the general tide of events. In that context, extraordinary failures will be tolerated. Thus the assertion by Herbert Simon in 1957[9] that a digital computer would be the world's chess champion within ten years, and that in the same time period, a digital computer would discover and prove an important mathematical theorem and that most psychological theories would have the form of computer programs by 1967, gave dramatic examples of over-optimism but did not curb the missionary zeal of the AI community in the United States. Similar underachievements in alternative systems would have been judged far more harshly. I personally therefore, never felt that the project would constitute a direct challenge to the existing systems, but rather it should be viewed as a form of technological agitprop around which new ideas could be explored and which would provide a sound foothold from which to advance further. The ultimate emergence of the ESPRIT project for human-centred CIM represented such an advance.

First Practical Moves

There can be little doubt that Howard Rosenbrock's reputation assisted enormously in getting the funding in the first instance, and also helped provide a credibility for the project as a whole. Here was an engineer with a successful track record as an R&D manager in industry, and an international reputation for his academic work in control systems theory. Furthermore, it meant that the proposal for alternative systems was coming *from within the tradition of science and technology* rather than a negative and destructive attack on those traditions from without. The Steering Committee was unique, at least in the British context, in involving a wide range of disciplines at the design stage. The work of the Committee was for me stimulating, but also problematic. Firstly, it was clear that social scientists and engineers use quite different cognitive maps, modes of operation and expression. It became evident that much further work had to be done before an adequate framework could be arrived at in which the two traditions could collaborate. The issue really required study in its own right. Even within the engineering profession itself, there are now serious discontinuities between its own internal specialisms. Add to this 'outside professions' and the problems increase dramatically. Yet the sheer complexity of modern systems and the pervasive nature of their multiplier effects, necessitates multidisciplinary design teams capable of a holistic view. It was also clear that it is notoriously difficult to transform sociological, ideological and political critiques into a series of technical solutions.

Furthermore, it is almost unheard of in Britain to encounter an engineer who has seriously studied social science, although in Germany and Scandinavia this would be a less rare occurrence. In Britain 'social effects' seems to be more rooted in sociology than in social sciences, as in Germany where there are often stronger links with industrial development. The sociological tradition tends to be strongly influenced by the 'critical studies' approach of the sixties, which appears to hold that a well structured critique is an end in itself. From an engineering point of view, the definition of a problem, in the sense of design methodology[10] is usually regarded as an important step in defining a solution. Engineers tend to be solution led and sociologists critique centred. Engineers are often driven by the tyranny of the immediate, whereas the social scientist may take a more timeless approach. If a functioning system is to be delivered as the culmination of a three year project, this may dictate that the hardware has to be purchased in the first six months, from the engineer's project scheduling point of view.

The social scientist, who is dealing with less project specific concepts (i.e. more universal ones such as alienation), may protest with some justification that by selecting the hardware at such an early stage, options are already being closed off before their potential has been adequately explored. The engineer is compelled to converge early, the social scientist will wish to diverge to a later stage in the project. This can be incorrectly interpreted as the engineer being entirely 'positive' and wishing to get on with things, and the social scientist as 'negative' by being indecisive and procrastinating. To say so, is not to condone the kind of academic rambling which turns working groups into talking groups. These are real practical problems which caused severe tension in the ESPRIT project (see below), and which must be faced and resolved if there is to be a symbiotic relationship with multidisciplinary design teams.

Issues of Status, Skill and Uncertainty

There was the problem of the user involvement, and this has been dealt with elsewhere in this book. I wish to refer only to a cultural dimension of this. In most sections of British Industry it is held to be 'progress' if one moves from the shop floor into a white collar job, however trivial. Highly skilled prototype fitters or toolmakers, who are essentially designing by doing, used to regard it as progress to become a detail draughtsman doing quite trivial and undemanding work. However, it was 'in the office'. I found it significant that some of the shop floor workers from Rolls Royce regarded the introduction of new technologies, and the button pushing roles it implied, as progress because, as one of them put it 'We will now become white collar technicians'. This would not be the case to the same extent in West Germany, where practical engineering workshop skills are highly valued, and do provide the basis, for those who wish it, to progress through the engineering profession. Yet having achieved those higher levels they will not seek to deny the significance of workshop and production engineering skills. Several well known German professors will proudly proclaim as the first item of their CV that they have served an apprenticeship at Mercedes or wherever.

Innovative projects of the UMIST kind, when completed, inevitably leave in their wake tantalising hindsights, heightened retrospective logic and a sense of opportunities inadequately explored and exploited. There was what I regard as a missed opportunity. In retrospect I believe that we rejected, without adequate discussion or thought, the possibility of innovative tactile inputs/outputs in the system. This became known in the Committee as 'keeping the handles', which seemed to trivialise what might have been possible. For example, we should have explored more deeply the potential of analogical systems of the variety developed by Professor Gossard at the MIT.[11] The range of inputs which contribute to the totality that is human competence includes certainly intellectual input, but also includes all of that acquired through our entire range of senses. Dreyfus in his book *What computers can't do* asserts that computers will never be able to think because they do not have bodies. He further asserts that intelligence requires understanding and a background of common sense 'that adult human beings have by virtue of having bodies, interacting skilfully with the material world, and being trained into a culture'.[12]

Perhaps we do not think adequately of the body as a 'synergistic system',[13] possessed of vast information processing and restructuring competences which provide 'a ready made system of equivalents and transpositions from one sense to another'.[14]

Udo Blum of the German Metalworkers' Union, the university qualified engineer who is also a skilled manual worker, still feels that this aspect of the problem should be further explored to provide greater tactile feedback from the machine to permit the skilled workers to program it in a manner in which they *conceptualise* the machining process and also for the worker to 'program' his or her muscular and nervous system, in the wider sense in which Dreyfus poses the issue.

The project highlighted concerns I have had for many years about a number of fundamental design assumptions and notions of 'scientifically designed systems'. Good design is generally held to mean the reduction, or where possible the elimination, of uncertainty. But if we perceive the human being in the system as constituting an uncertainty, it follows, with some form of Jesuitical logic, that good design is about marginalising this human input. It is, however, noteworthy that in workerless factories, or those where human intervention is reduced to a minimum, the resultant systems are incredibly 'brittle' and vulnerable to disturbance. If one part of the system goes down, the high level of synchronisation is turned into its opposite and the entire system goes down. They tend to lack systems robustness. However, if there are highly skilled human beings, they are good at coping with uncertainties and provide for greater overall systems robustness.

Recent comparisons of downtime of advanced manufacturing systems in Britain and Germany, demonstrated that the downtime was significantly less in German systems and an explanation was advanced, which I hold to be true, that the Germans use skilled workers who will have had an apprenticeship or its equivalent and can anticipate systems problems and correct them in advance and in the event of a systems failure are capable of quickly taking remedial action. In Britain, the workers used are frequently products of Mickey Mouse training schemes and are little better than machine minders. This does mean that in the UMIST system there is retained the 'uncertainty' of a significant human input; and this is an advantage.

The project also highlighted for me, difficulties with the notion of 'scientifically designed' systems. Such systems are generally required to display the three predominant characteristics of Western science, namely predictability, repeatability and mathematical quantifiability. This by definition precludes intuition, subjective judgement, fuzzy reasoning and tacit knowledge and leaves little space for human intervention. In my view, a truly symbiotic system would require acceptance of human skills and attributes and would provide space for human knowledge, wisdom and action, rather than an obsession with data and information (see Fig. 10.1).

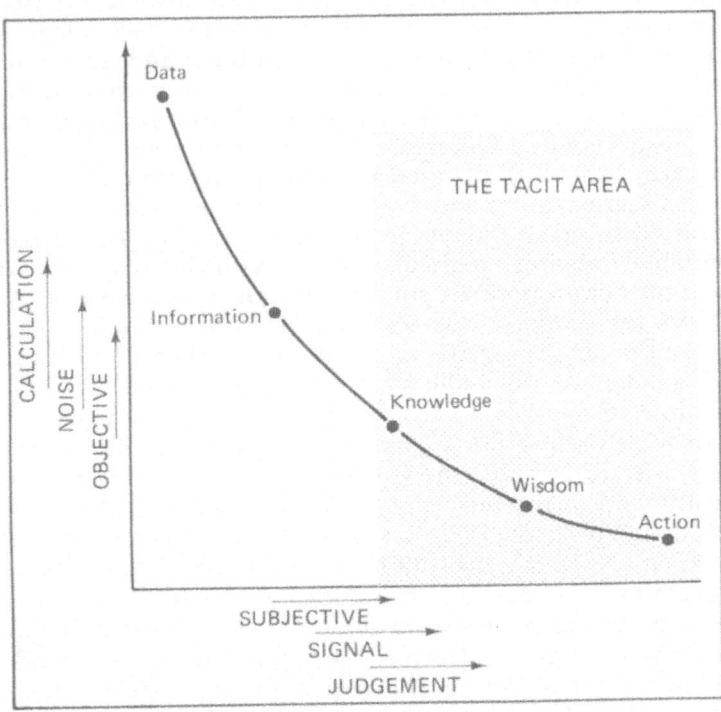

Fig. 10.1. The tacit area.

Human-centred Capabilities

A year into the project it was evident that even if one could demonstrate the human-centred capabilities at the level of one lathe, the sceptics would argue that, in the context of a manufacturing technology as a whole, it would not be viable – e.g. it would be incompatible with CAD systems and an advanced CAP environment. It was also clear that significant further resources would be required to progress the concept more generally. The project was by that time already attracting considerable international attention, and the early equipment had been inspected by researchers from the Metal Workers' Union in West Germany, the University of Bremen and the Technical University of Denmark. Out of these discussions arose a proposal to apply to the EEC for an ESPRIT grant to design and build the world's first human-centred computer-integrated manufacturing system. The proposal to Brussels asserted unashamedly that science and technology should be regarded as part of culture, and just as cultures produced different languages, different music and different literature, why should there not be different forms of technology. Should we not for example, develop a technology which reflects European cultural and social expectations. Granted such expectations have often more to do with the rhetoric than the reality, but at least in the rhetoric there is our concern about motivation, the dignity of the individual, free will, a sense of quality and support for creativity.

Furthermore, it was held that Europe had a cultural base diverse and rich enough, and an economic base large enough, to develop technological infrastructures and products which reflected its cultural and economic requirements. Such developments might differ significantly from those emanating from the United States or even Japan. The initial proposal was prepared with the Greater London Enterprise Board as the Prime Contractor together with two partners from the United Kingdom, three from Denmark and three from West Germany. In the event, the first proposal was not accepted but it attracted so much interest in Brussels, and was also frankly the basis of considerable lobbying – particularly from West Germany – that it was suggested that a further proposal be submitted in the following year. At the beginning of 1986, the proposal was accepted for a £3.8 million project and was joined at that time by a powerful German industrial partner, Krupp Atlas Elektronik. I do believe it would have been very difficult to have obtained this EEC project[15] had it not been for the sound basis of work and demonstrable results which were already emanating from the UMIST project.

Once the project was underway, there was a growing interest in it worldwide, and at the end of the first year the project was joined by two powerful British industrial partners, Rolls Royce and BICC. The early stage of the project reflected many of the difficulties encountered in the UMIST project, and the experience of dealing with them was helpful. Tensions between social scientists and engineers were evident during the early stages, particularly in the CAM work group. The CAD group is based in Denmark, the CAP group in Germany and the CAM group in Britain. Synchronising their work is proving to be difficult but stimulating. The commitment to the notion of human-centredness is at varying levels and assumes several different forms. The human-centred CAM cell is due to be at Rolls Royce in Leavesden in October 1988, and the integration of the CAP system will take place in Bremen around the same time. The installation at Rolls Royce will be available to small and medium sized companies and other interested parties on an agreed basis for 50% of the operating costs for two years from October 1988. This will provide a unique opportunity for familiarisation with such systems and the production of short batches so that realistic comparisons can be made with the conventional system. The project is now funded to a level of £5 million, and subject to half-yearly reviews will run for three years to May 1989. It is the intention to provide a system in which the human being will handle the qualitative subjective elements. This will be done in such a fashion as to enhance human skill and ability and provide those involved with powerful tools.[16] A full description of the project is provided in Chapter 11.

The project represents an important outgrowth of the UMIST project. It does not necessarily mean, however, that the idea of human-centredness is now established and will expand. It may be viewed by some merely as a transitional technology, useful only until more powerful machine-based systems have been developed. There is possibly the challenge that the developing technologies will endanger its survival. For example, developments in computer-aided design – in particular, solid modelling capability – may make it attractive for employers to down load the machining data straight on to the machines, so by-passing the skilled workers on the shop floor. There is also the less obvious but more insidious danger that the term 'human-centred' will be applied to a range of systems which display nothing more than a slightly higher degree of user friendliness and will not fundamentally change systems design. This could actually damage the concept of human-centredness in the long term. One could imagine a new sales gimmick in adverts proudly proclaiming some trivially modified system to be 'Human-Centred and User Friendly'!

Future Development of Human-centredness

Trade unions world-wide are increasingly interested in human-centred systems as an alternative to the deskilling effects of many modern systems. The International Metalworkers' Federation, which represents some 170 trade unions in over sixty countries, is circulating a report prepared by the writer which includes a section on this topic.[17] The 2.8-million-strong Metal Workers' Union in West Germany has facilitated the publication in German of a collection of Howard Rosenbrock's papers, and separately has produced a booklet in German and English on the need to retain human skill.[18]

In order to underpin these ideas at an academic level and establish some theoretical basis, and to facilitate postgraduate programmes in these areas, a new journal entitled *AI & Society: The Journal of Human and Machine Intelligence* has been founded and has on its Editorial Board researchers and academics from most of the industrial countries.[19]

Discussions are at an advanced stage in the plan to establish an 'International Institute for Multidisciplinary Research in Human-centred Systems'. This will be an institute 'without walls' and will link researchers world wide. Its aims include the following:

To facilitate debates, seminars and dialogues on key intellectual and philosophical issues in this area

To provide an international forum for these debates and a network of those active in these areas who share a concern about the social effects of technology and are interested to restructure technology to serve the long term betterment of humanity.[20]

The UMIST project can correctly be seen as a point of departure for these more recent developments. It was a valuable aperture through which to view the problems of a project of this kind, and the difficulties in practice of transforming theoretical concepts and critiques into hardware and software.

I have suggested in my book *Architect or bee?*[21] that we are now at a unique historical turning point in which decisions that we make in respect of the new technologies will have a profound effect upon the way that we relate to each other, to our work and even to nature itself. I suggest that a similar historical turning point existed between the twelfth and sixteenth centuries. The debate culminated in Dürer suggesting that it would be possible to develop forms of mathematics which were 'as amenable to the human spirit as natural language'.[22] That opportunity was sadly lost, although the debate went on in one form or another for some 120 years. Given the rate of technological change today I would suggest that we have perhaps 20 years in which to demonstrate convincingly what the alternatives might be like. Otherwise, an infrastructure of machines, systems design philosophy and educational forms which marginalise human intelligence and skill will continue to evolve. The UMIST project has helped to focus attention on constructive alternatives, and has excited the imagination of many young and able researchers who, it is hoped, will carry these ideas forward.

Perhaps Sarson is correct when he suggests 'Human Centredness might even become the bandwagon of the 1990s'.[23]

Acknowledgement. SERC/SSRC are to be thanked and congratulated for the courage and wisdom in supporting such an unusual and challenging project.

References

1. N. Wiener, The human use of human beings, 1967, p. 254 (Avon Books – Discus edition).
2. Cf. various works by J.D. Bernal: e.g. The world, the flesh and the devil: an enquiry into the future of the three enemies of the rational soul (London 1922); The social function of science (London 1939 and Cambridge Mass, 1967 edition); The freedom of necessity (London 1949); Science in history (in 4 vols. Penguin Books 1965).
   A fascinating account of the lives and times of British Socialist and Communist Scientists, including Bernal, is to be found in G. Werskey, The visible college, 1978 (Holt, Rinehart and Winston).
3. See ref. 1.
4. Cited in S. Bodington, Science and social action, 1978, p. 20 (Allison and Busby).
5. J. Lobell, Design and the powerful logics of the mind's deep structures, Design Research Society Journal, vol. 9(2), pp. 122–129.

6. G. Friedrichs (editor), Aufgabe Zukunft: Qualität des Lebens, 1972 (Europaische Verlagsanstalt).
7. M. Cooley, L. Friberg and C. Sjoberg, Alternativ Produktion, 1978 (Liberforlag).
8. H. Wainwright and D. Elliott, The Lucas plan: a new trade unionism in the making, 1982 (Allison and Busby).
9. H. Simon cited in H. Dreyfus What computers can't do: the limits of artificial intelligence, revised edition, 1979, pp. 81–82 (Harper & Row).
10. J.C. Jones, Design methods: seeds of human futures, 1970 (John Wiley).
11. D. Gossard, Analogical part programming with interactive graphics, PhD thesis, 1975 (MIT Press).
12. Dreyfus, see ref. 9.
13. M. Merleau-Ponty, Phenomenology of perception, 1962, p. 234 (Routledge & Kegan Paul).
14. See ref. 13, p. 235.
15. ESPRIT project 1217, Human Centred CIM. Details from Greater London Enterprise, London SE1.
16. M. Cooley and S. Crampton, Criteria for human centred systems in ESPRIT CIM, 1987 (North Holland).
17. Technology, trade unions and human needs (a 50 page report available free in any of the following languages: English, Finnish, French, German, Italian, Japanese, Spanish, Swedish), 1984 (International Metal Workers Federation, 54 bis Route des Acacias, 1227 Geneva, Switzerland).
18. Available from I.G. Metall, Frankfurt, Germany.
19. AI & Society: The Journal of Human and Machine Intelligence, July 1987, vol. 1(1).
20. Details from: M. Cooley, Thatcham Lodge, 95, Sussex Place, Slough, Berks., England.
21. M. Cooley, Architect or bee?: The human price of technology, 1987 (Chatto & Windus/The Hogarth Press).
22. See ref. 21, pp. 55–56.
23. R. Sarson, A technology aiming for the best from hand and screen. The Guardian, 4 February 1988.

Chapter 11

The ESPRIT Project

Shaun Murphy

Introduction and Summary

This chapter gives an overview of ESPRIT project 1217 (1199), 'Human-centred CIM Systems', at the beginning of the final year of the three-year project. The object of the project is to produce human-centred CIM components, in which human skill and its application are optimised in harmony with leading-edge computerised manufacturing technology.

This kind of technology is particularly appropriate for small and medium sized batch manufacturing: and the project's industrial partners are seeking new solutions to competitive pressures which demand shorter delivery times, lower inventories, and better quality control.

The relationship between this ESPRIT project and the UMIST project has already been described by Mike Cooley in Chapter 10.

The chapter begins with a description of the qualities of human-centred technologies, and this is followed by an idealised scenario of a factory of the future, organised and equipped according to the human-centred philosophy. These two sections taken together provide a picture of the thinking from which the work of the project has developed.

The description of the work focuses on the project's three demonstration sites, as well as specific human-centred CAD, CAM, and CAP products. The BICC demonstration site at Sealectro Limited, and the Rolls Royce demonstration site at the Leavesden plant, are both industrial demonstrations, and the BITZ demonstration site on the campus of the University of Bremen is an experimental site at which further developmental work can take place. The CAP product is a shop-floor monitor and controller which provides production island operators with scheduling and production management support. The CAM product is a lathe controller with extensive colour graphics facilities, which enables turners efficiently to produce part-programs on the shop-floor. The CAD product is a sketching module on which designs can be sketched in pencil and electronically recorded at the same time, and which will act as an effective means of visual communication between designers and manufacturing staff.

A glossary of some abbreviations and technical terms will be found at the end of the chapter.

The Meaning of Human-centredness

Human-centredness is not a concept which can easily be defined in a few lines, but, in general, a human-centred technology is one which extends human skill and its applications.

Discussion is still going on within the project about the nature of human-centredness, and the concept will be developed further in the coming year. Individuals place different emphasis on different aspects and this section represents the author's current views.

In the field of manufacturing, Rauner, Rasmussen and Corbett[1] have pointed out that a human-centred CIM system must be developed in such a way that each of the following three areas are thoroughly analysed.

First, the technology itself, that is the hardware and software, must be designed in such a way that the synergy between human skill and computer power is optimised. Shop-floor part-programming of a machine tool is a good example of this.

Secondly, work within the factory must be organised so that in all areas people are able to apply their skills. In particular, people on the shop-floor should have a significant degree of responsibility for production activities.

Thirdly, individual skill and competence should be increased through a balanced combination of learning by doing and formal training and education.

From the point of view of the individual, Ainger and Murphy[2] have described a number of human-centred qualities:

1. *Retention and enhancement of existing skills:* A human-centred system is one in which traditional design and manufacturing skills are used to their fullest extent. Whilst a system may be used by people with varying degrees of skills, it will be most productive when used by skilled people. The nature of skills will change with time, but at present, for example, the skills appropriate for a turning cell are those of a turner who has learned the trade on traditional lathes, and who has some experience of NC machines.

2. *Extension of operator choice and control:* The person should control the machine rather than the machine controlling the person. In particular, the timing of actions and events should be decided by the operator. The operator should be able to move freely around the shop-floor, and outside it if necessary. People should be able to communicate freely with each other, both formally and informally, and both electronically and face to face. Within the framework of production targets the operators should be personally responsible for organising their work in such a way as to achieve those targets. Choice and personal control in these areas will regulate the amount of stress to which people are subjected.

3. *Subdivision of the work should be minimised:* The range of activities for which each operator is responsible should be maximised. In this sense human assertiveness is the antithesis of Tayloristic scientific management. The aim of Taylorism is to sub-divide work as finely as possible in order that each minute element can be prescribed exactly and performed by someone with as little intelligence and training as possible (see Chapter 1, p.2).

4. *Operator knowledge of the whole production process should be increased:* Each person should have a general knowledge of the whole production process, and should have the opportunity to comment on any aspect of it. The forms of human-centred systems should be such that knowledge of the whole production process is easily acquired, and communication between people working at different points in the system should be facilitated. Communication should be both formal and informal. For example, it may be that communication between designers and turners may be more productive if it is a conversation over a drawing rather than around an electronic screen.

5. *Physical ergonomic factors should be considered:* These ergonomic factors include, for example, the physical arrangement of equipment, design of keyboards, and user friendly aspects of software. These factors as well as, for example, safety considerations are very important. A system could hardly be described as human-centred if it was not engineered adequately; however, these factors do not in themselves constitute fundamental human-centred qualities.

The combination of the application of Taylor's philosophy of scientific management with the development of computers produced a drive towards centralising decision-making. The effects of this can be seen in most of today's factories in the fields of design, manufacture and planning, particularly the latter.

A conscious and thorough-going policy of pushing decision-making down to the lowest possible level of the management hierarchy, and as near as possible to the point of production (in most companies this amounts to the same thing), would be a major step towards human-centred manufacturing. Over a period of time this policy, if pursued seriously, would influence the design of the technology, give substantial responsibility to shop-floor operators, and increase levels of skill and qualification.

The Factory of the Future

The ideas of human-centred manufacturing can be presented more vividly by extending this rather bare description of the concept to a portrayal of what a human-centred factory would be like.

This factory is presented in some detail, but it is not suggested that there is only one type of ideal human-centred factory.

This ideal factory bears little resemblance to the great majority of factories in the world today, although some of the components of the scenario are present within some companies, particularly in West Germany, that have successfully adopted semi-autonomous group working practices.

This description or scenario of the human-centred factory of the future was developed by the International Social Science Group within the ESPRIT project and is outlined in more detail by Rauner, Rasmussen and Corbett[3] from which the following is taken.

The human-centred factory of the future focuses on small to medium batch production and contains flexible production islands incorporating both highly automated and manual-intensive machines. 'Flexibility' here is taken to mean the ability to adapt to changes of product design, batch size, and machining processes and sequences in order to meet changing market demands and customer requirements. To meet these demands, the human-centred technology must be integrated with, and complemented by, an equally flexible organisation of work.

Within the factory, the technology and the organisational structures around it have been designed and developed through an evolutionary process in which all relevant personnel have collaborated and participated. The objective of this process has always been to consider economic, technical and social considerations from the very onset of design in order to optimise technical efficiency and flexibility at the same time as optimising the quality of working life of factory personnel. (In this context the phrase 'quality of working life' is used in a more fundamental sense than that implied by issues such as the length of tea breaks.) In addition to the normal forms of technical and computing expertise associated with CIM system design, the design process has involved collaboration with other people, from line management, users, trade union representatives, suppliers, and other relevant experts, including social scientists and human factors experts.

Range of Products

The factory produces a wide range of products with a certain common core. These products include 'fixed' products presented to customers in catalogues, products adapted to customer requirements and customer-specified products. Relationships with customers are direct to allow the evaluation and improvement of the use-value of the products. In this way the factory can develop its product range in new directions as expertise in design and manufacture (and customer requirements) develop.

The Factory

The factory is made up of three sectors: coordination (which includes customer liaison, sales, MRP and production planning), design and manufacture. These are organised by product and not by function to maximise flexibility. In other words, instead of dividing labour, work is organised by dividing products and product orders.

This is achieved by extending the concept of group technology into a general organisation concept rather than a single technique. The management structure is based on the idea of a 'flat hierarchy'; that is, there are the minimum number of hierarchical levels in the factory. By restricting the length of the managerial chain, people are more in touch with the wider aspects of factory life, the management are more reactive and flexible, and the quality of decision-making is consistently high.

The factory's product range is divided into part families, i.e. groups of the products with similar characteristics, and the production islands are responsible for a particular sub-group of these part families. The production islands have the task of producing components and complete products in so far as possible from raw materials. All necessary human, technical and material resources are therefore concentrated within the production island.

Design and manufacture are organised around these part families so that each product group in manufacturing has a corresponding group in the design section of the factory. The designing process is split up according to product families, or their parts, so that the designers perform the whole design process comprising tasks such as determining functional structures and dimensions, geometric modelling and designing new products. Thus, two skill-centred production sub-groups have been formed equipped with local computer assistance and connected by electronic data exchange: namely, production and design islands.

These two islands are interlinked by the basic components of CIM architecture. These are:

1. A common data base with which all functional programs may interact
2. A data highway to link sub-systems
3. Data exchange interfaces, to integrate sub-systems, customer orders and material supply

Unlike the centralised system architectures commonly associated with Taylorist work structuring, these components reflect a CIM environment that is integrated in terms of information rather than control. In other words, instead of formalising and incorporating almost all production knowledge and work planning into the computer system, the computer system serves as an integrated information system. Although the computer controls routine operations, its decisions are subject to amendment or overruling by the personnel, who do not lose computer support when they exercise this supervisory control. The planning of work activities is left to the island personnel, who use their knowledge and skill to optimise island performance, with the computer system providing them with accurate information and simulations to support decision-making.

The interlinking of island activities is organised by the coordination department, which is responsible for maintaining the flow of information between islands, and for distributing tasks to the relevant design and production islands as well as initiating new product groups, and for implementing new island configurations.

The presence of a coordination department does not mean that planning and execution are distinct functions. Orders being produced in a production island do not require detailed process planning information from the coordination department. It is sufficient only to specify the completion date for the part or product and not the sequence of operations. This is left to the members of the relevant production island to decide. Typically, the production scheduling (which indicates work to be completed over a period of one week) is carried out by the coordination department, whilst the production sequencing and monitoring is the responsibility of the production island personnel.

Furthermore, there is extensive collaboration between all three sectors of the factory by means of electronic data exchange, as well as informal and face-to-face communication. In a way it is perhaps more accurate to describe the factory system architecture as computer-aided integration, rather than computer-integrated, manufacture (CAIM rather than CIM) as everyone is aware that verbal communication and debate may, in many instances, be a more effective medium for the exchange of information and ideas.

For example, new interdepartmental groups are formed whenever a new product or product family is to be developed. In this way, the product will be adapted, as far as possible, to suit the manufacturing capabilities of the production island from the earliest stages of the product design process. This strategy of designing for manufacturability includes the consideration of how product quality may be affected by such methods, and regular feedback is encouraged from the purchasers and users of the factory's products.

On a more day-to-day basis, the interactions between coordination, design and manufacturing enable all personnel to see their work from the perspective of its wider implications for the overall factory, and also to be involved in a wide range of decision-making activities. For example, once a product delivery date has been agreed by coordination, design and production personnel, any unforeseen problems arising from within the production island which threaten punctual completion are dealt with by the production island personnel. If the boundary manager for the island (see below) detects that re-scheduling within the island will not solve the problem he or she may then attempt a form of load sharing with other islands before alerting the coordination department that delivery scheduling requires reprocessing. These re-scheduling activities are carried out with the aid of the factory-wide computer-aided planning system, with overall management being the responsibility of the coordination department personnel.

The Production Island

The production island has four main tasks:

1. Production planning (in collaboration with the design group)
2. Production (machining and assembly, plus equipment maintenance and quality assurance)
3. Product and methods development (and including technology investment planning)
4. Qualification and training development

The production island work group comprises a number of 'electro-mechanical craftspeople' who possess skills in machining, assembly, machine setting, NC program generation and editing, inventory management, quality assurance, machine maintenance, and work scheduling and planning. Although all group members possess these skills, some members have developed a more specialised interest in certain tasks and functions (e.g. hardware maintenance). Unlike the rigid division of labour associated with the more Taylorist organisational structures, the island has an 'organic' division of labour in which individual tasks and skills overlap to allow group problem solving and flexibility in personnel–machine assignment.

Although the group is free to organise and supervise its own behaviour, the internal division of labour follows two guiding principles:

1. The division is always horizontal rather than vertical (e.g. planning and task execution are rarely divided among group members; each group member being responsible for the setting, programming and machining functions for any given machine)
2. It shall always be possible for every member to experience and develop the relation between quality of product and quality of production (e.g. all personnel participate in the four categories of tasks outlined above)

One person in the group has the responsibility for managing the boundaries of the island. This worker negotiates with the coordination department on which jobs to initiate and when they are to be delivered. She, or he, is also responsible for the liaison between the island and stores and the materials supply department to ensure that the island always has sufficient resources to undertake the work in progress. This leaves the other group members free to concentrate on production (although they will be involved in briefing the boundary manager). The boundary manager coordinates distribution of tasks for others in the group.

The Design Island

Personnel working within the design island are organised along the same lines as those within the production island in that the division of labour is based on product responsibility rather than a functional specialisation. Designers, therefore, tend to work for a particular island or group of islands.

There are a number of design teams within the design island, each team being responsible for a particular family of products and for nominating people to participate in the interdepartmental product groups which are formed whenever a new product is to be designed and manufactured. Participation in these groups enables designers to increase their knowledge and understanding of production and coordination within the wider factory.

In line with the 'organic' division-of-labour philosophy applied in the production islands, designers tend not to be dedicated users of one particular workstation, but use most of the technical equipment within the design island during their work.

The CAD workstations within the design island allow traditional drafting and designing skills to develop as the technology develops. For example, on many of the workstations in the island, the designer's initial draft or sketch is transmitted directly to the CAD software. This can be achieved in one of three ways, the choice being left to the users' own preferred method of working; drawing data can be simultaneously recorded while the designer works at a computerised drawing board using traditional design aids (ruler, etc.), or the completed drawing or sketch can be digitised, or the geometry may be directly entered by means of more traditional data entry techniques (e.g. soft key menus).

ESPRIT Project 1217 (1199)

The German, British and Danish partners began work on the three-year project in May 1986. A list of participating organisations is shown in the Appendix to this chapter. The German partners are carrying out the computer-aided planning (CAP) work; the British are concentrating on the computer-aided manufacturing (CAM) developments; and the Danes are performing the computer-aided design (CAD) work. The project is due to run for three years, and the total budget is just under £5 million.

This description of the work was written at the beginning of the final year, and focuses on the three demonstration sites, and the three principal human-centred products.

The BICC and Rolls Royce demonstrations are both at industrial sites, and the BITZ demonstration, on the campus of the University of Bremen, will allow further development to take place free from production constraints.

The three products are the shop-floor monitoring and control workstation, the sketching module, and the human-centred lathe controller. Whilst it would be technically possible to exhibit all three products at each of the demonstration sites, constraints of time and money will probably limit the demonstration of the first two to the BITZ site, and the latter to the Rolls Royce site.

Project Objectives

The overall object of the project is to produce human-centred CIM building blocks, in which people are given responsibility for those tasks best done using human skills, whilst the use of computerised systems in support of the shop-floor operators is optimised.

The primary objectives of the project are:

1. Establish criteria for the design of human-centred systems
2. Establish their economic and commercial competitiveness
3. Achieve a higher level of flexibility and robustness against unforeseen events than is possible with conventional systems
4. Achieve a lower throughput time from design to finished components than is possible in conventional automated systems
5. Achieve a better working environment and working practices by developing the synergy arising from man–machine interaction
6. Assess and further develop the effectiveness of man–machine interaction
7. Establish the means by which the training of operators can be most successfully carried out
8. Design flexibility into the modules, to facilitate portability, and re-configuration with other systems
9. By the success of the demonstrations, show that there is a better means of organising production, which is specially suited to European conditions

CIM Design Considerations

It is a fundamental tenet of human-centred CIM systems design and development that the specification of the technological system cannot be seen in isolation from the social system within which it will operate. One of the problems associated with the design of conventional manufacturing technology is that it is dominated by technical specialists who are driven by technical considerations. A number of problems result from this technically led approach:

1. Early technical design choices may constrain the design of subsequent social aspects, and may lead to sub-optimal system design
2. The technical domination of the design process may mean that a coherent plan for the organisation of work around the technology is not developed
3. If social aspects such as human–machine interface design, job design, training, management structure and coordination are ineffective, then a system will not perform to its potential

In this project technical and social aspects have been considered from the beginning of design in order to harness the skills, knowledge and flexibility of system users and support personnel to the full, both during design and after the system is implemented in an organisation[4]. The specifications for the various components of human-centred CIM explicitly include social and organisational requirements.

The BITZ Demonstration Site

The Bremer Innovations und Technologiezentrum (BITZ) was founded jointly by the University Senate and the Chamber of Commerce in Bremen. It is situated on the campus of the university, and a demonstration production island is being constructed here which will highlight the German CAP developments, and the Danish sketching module.

The demonstration has three purposes:

1. To be the basis for the development, and testing of the human-centred CAP software, and the sketching module
2. To aid the integration of human-centred CIM components with each other and, where appropriate, with certain elements of conventional manufacturing technology
3. To act as an exemplary demonstration site for aspects of human-centred CIM

The layout of the production island is shown in Fig. 11.1. The island will consist of the following machines:

1   flame-cutting machine (SUAG)
1   saw

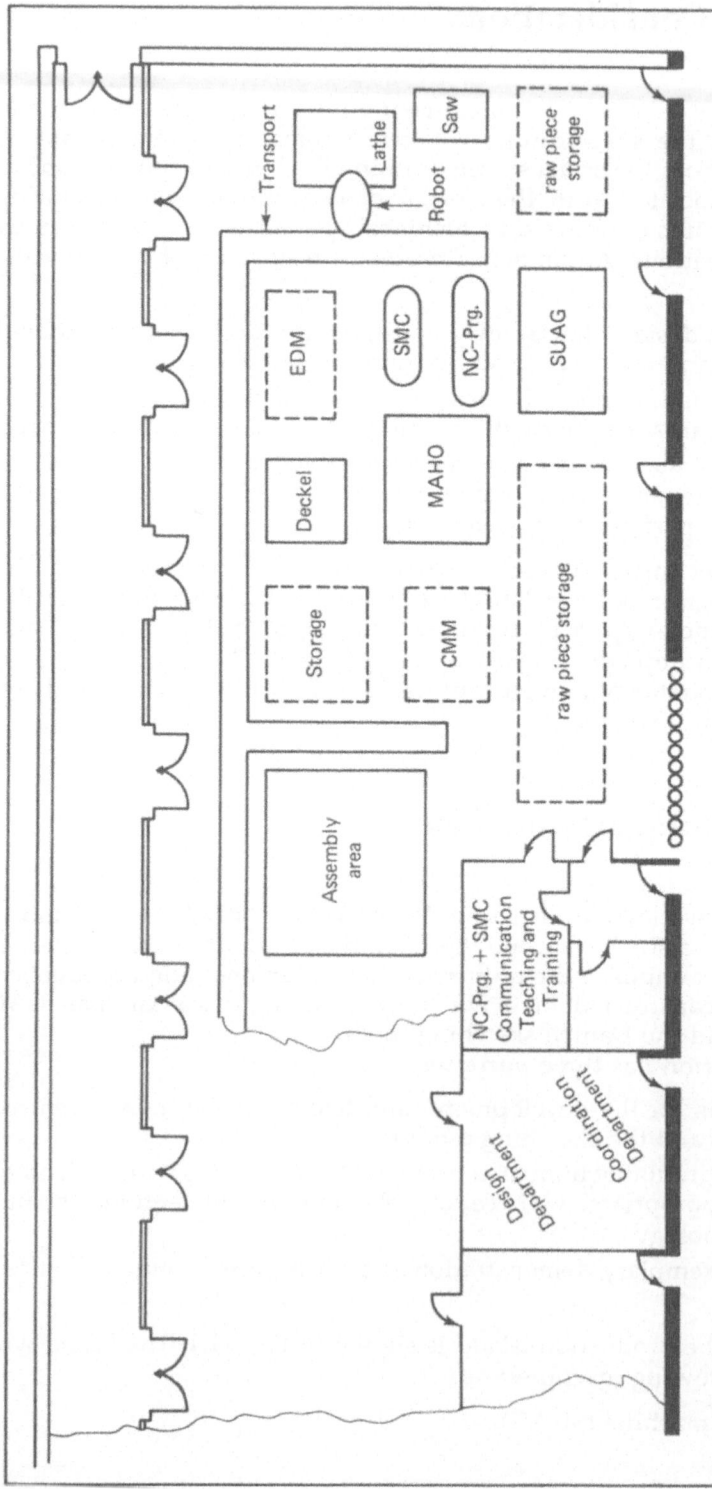

Fig. 11.1. Production island for metal cutting and mould manufacture (HC CIM, ESPRIT 1217 (1199)). CMM, computer-monitored manufacture; for other abbreviations, see the text.

1   lathe
1   3-axis drilling milling machine (Deckel FP3NC)
1   5-axis milling machine (MAHO 700 c/A)
1   robot
1   electric discharge machine (EDM)
1   measuring machine

There is an area for manual assembly in the island as well as a separate room for discussion and teaching and training.

The cutting, turning, drilling and milling of simple components can be performed on different kinds of materials such as metal, wood and plastic. Thus the machine configuration covers the essential requirements of typical processes in the machine tool industry.

The Shop-floor Monitor and Controller

The main thrust of the human-centred CAP work is the development of a workstation to assist with scheduling within a production island. This SMC workstation will make visible material flow and product status within the island, and will provide information about the change of outcome when a change in the sequence of jobs is proposed. As with the human-centred lathe controller, the philosophy is that cell operators receive scheduling suggestions and monitoring information from the SMC, but scheduling decisions remain with the operators, who can override computer-generated schedules.

The system will provide the island operators with the following facilities:

1. Surveys in overview of the workload allocated for the time up to the planning horizon. Users will be able to specify the way in which the overview is presented (e.g. breakdown by order, by type of machining operation, by date, or by machine).
2. A means of setting the priorities by which the job steps are allocated to machines in the cell.
3. A means of recording the non-availability of resources for performing certain operations.
4. A means of ascertaining which orders cannot be completed by the required due date.
5. A means of ascertaining which resources (materials, tools, NC programs, etc.) are required to perform a particular order or machining operation and whether these will be available at a given time.
6. A manual override facility to the automatic resource allocation which will 'freeze' a particular operation to a specified point in time.
7. A means of ascertaining the expected utilisation of a particular resource (e.g. tool, handling equipment, operator, machine).
8. A means of introducing operations not requested by the production-planning system (e.g. testing of NC programs, special orders for replacement workpieces, rush orders, maintenance operations, and equipment

repair) and ascertaining their effects on due dates for the required production schedule.

9. A means of ascertaining the stage of processing of the complete order to which a particular suborder belongs.

10. A means of informing other cells about missing resources which could be compensated for by excess capacities on other cells.

11. A means of testing whether an improved machining sequence to manufacture a particular part would lead to an improvement in machine utilisation.

12. A means of informing the production scheduling system of planned 'real' capacity several planning cycles in advance.

13. A means of recording particular successful sequencings which will be used for training less skilled operators, and a means of recording special situations which will be used to evaluate the effects of introducing new planning procedures.

14. A means of presenting a training programme in scheduling and sequencing to the operator.

15. Feedback of quality control data, with a view to improving the efficiency of the manufacturing process while maintaining the required quality.

16. A means of reconfiguring the island to include or exclude individual machines or workplaces.

17. Sending instructions to other sub-systems (e.g. storage and transportation) to provide resources (e.g. tools, blanks) at a specified time.

The software will be implemented on a workstation based on the Motorola 6800 chip with colour graphics facilities.

The Sketching Module

As with manufacturing, industrial design is a work culture in transition, partly based on traditions and partly on the changes introduced by commercial pressures and technological innovation.

Although design activity has become increasingly separated from the manufacturing process, many industrial designers are still recruited among craftsmen and, until now, have been an important connecting link between the more analytically oriented engineer and the more practically oriented craftsman.

The traditional design method is to make drafts and redrafts, either on different parts of a large piece of paper or on a series of tracings from the original sketch or layout. The effect of concentrating the geometric aspects of manufacture in a drawing is to give the designer a much greater 'perceptual span' than the craftsman had. Thus, the process of design-by-drawing can be seen as an accelerated version of craft evolution, with the freedom to change several parts at once, rather than only one at a time.

The drawing has three main functions in the design process as Cooley[5] mentions:

1. To help in the conceptual phase of design. Here a drawing enables a designer or draughtsman to visualise what he or she requires. It confirms feasibility, and aids the transmission and acceptance of ideas.
2. As a means of communicating the design intention to manufacturing, inspection, toolroom, assembly and other departments. Drawings indicate what has to be done to produce the product and make it work.
3. As a memory bank for what has been done. This allows the drawing or design to be modified, updated and reused.

The Danish group are developing a sketching module, the use of which will preserve these three functions of a drawing. The module will allow pencil sketches on paper to be produced, whilst at the same time storing the data electronically in such a form that it can be transmitted to a CAD system.

The sketching module can be used for both design work, and work preparation. 2-D or 2.5-D geometry can be input to CAD or CAM, and the ideal functionality is shown in Fig. 11.2. The sketching module makes it possible to use sketches, old drawings, or IGES-translated CAD drawings as inputs from which the design of a new component can be made. The geometry can be made and edited by either the designer or a member of the manufacturing staff and exchanged between them. Because the CAD entry system is the same for both, they can easily communicate and possibly change the design. The module will be easy to carry around if desired, and thus it makes informal personal communication easy, compared to the more stationary CAD systems. The module may improve learning possibilities. Since it is easy to learn how to use it, it may be an advantage to introduce the system to designers who are not familiar with CAD systems. On the other hand, persons very familiar with CAD systems but not with the work culture of the design process may be introduced gradually to this by using the electronic sketch block in cooperation with a more experienced designer. The present prototype is shown in Figs. 11.3 and 11.4.

In the beginning of the design process, the designer is expected to use sketches to express his or her ideas. These ideas can be drawn in shapes using various components on the sketching module and stored in the computer's idea bank. The idea bank can be distributed to others in the design team for further work. After deciding the shape of a part, it is drawn in the exact shape geometrically on the sketch block.

By this time the drawing can be transferred to the CAM area to be examined and modified. In this process the designer and production planner can work on the shape and use the designer's notes on tolerances, and surface finish to obtain a part which can be produced on the available machinery. The drawing is then transferred to a CAD system for additional information which completes the documentation of the drawing.

The BICC Demonstration Site

BICC's island will involve a wide span of human-centred activities across the organisation as a whole by embracing a range of trades, functions, and skills in an island manned by a multi-disciplinary team of individuals who will be provided with a full range of manufacturing support technology, but who will be using relatively standard manufacturing processes.

The main stimulus for the work has been that manufacturing lead time is currently longer than the market demand lead time, and hence manufacture has to be carried out against a forecast schedule rather than a firm schedule. This is not a problem for approximately 20% of the company's turnover

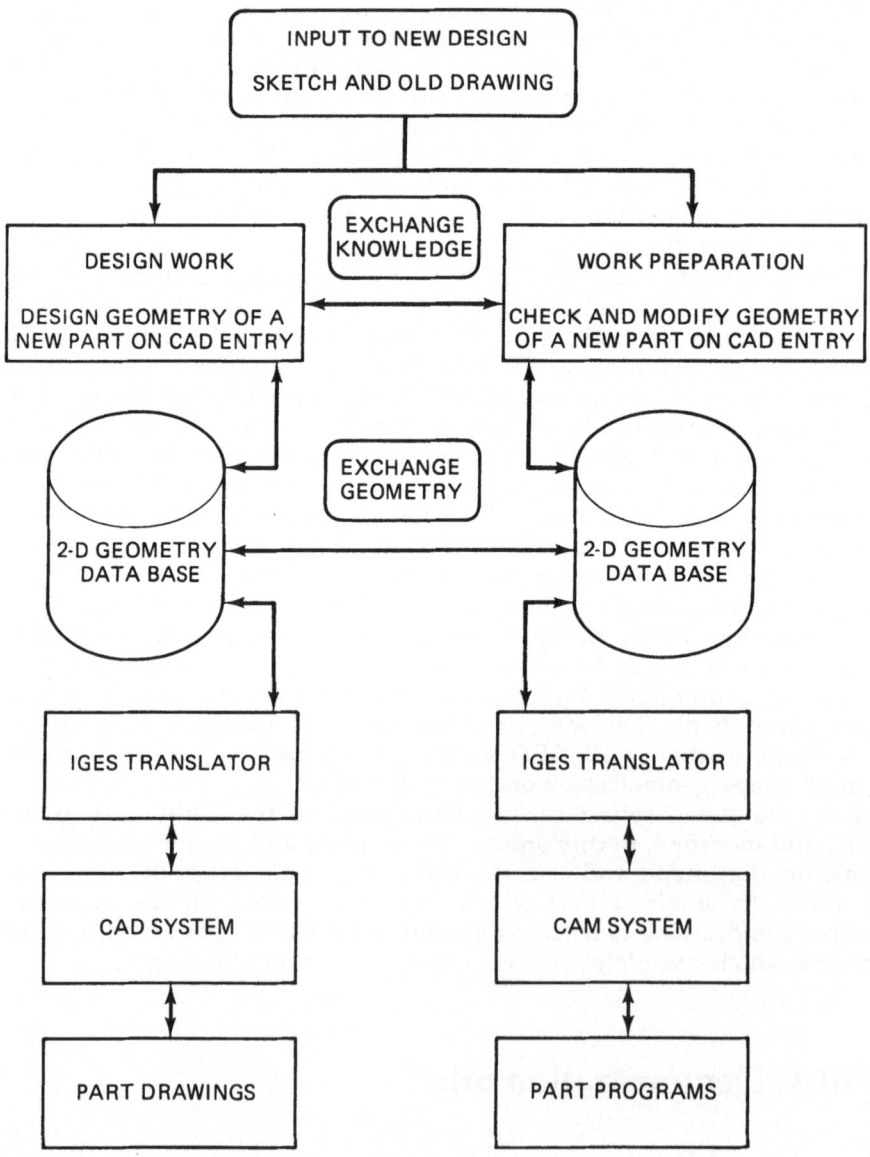

Fig. 11.2.   The ideal functionality of the suggested CAD entry medium.

Fig. 11.3. Sketch of the demonstration model.

Fig. 11.4. First prototype of the demonstration module.

which is for sufficiently high-volume products to allow a continuous rolling forecast to be maintained with supply from stock, but for the remaining 80% of the turnover several problems are created. These include losses due to obsolescence, high inventory levels, high management costs incurred in administering forecasts, and variable due date performance.

Factories have traditionally tended to be arranged upon a functional layout basis. In other words machines were grouped together on the basis of process similarities with products moving from one process area to another, covering a large distance in a quasi-random route pattern. The difficulties associated with tracking components in a situation like this have often led to increased investment in more sophisticated control systems, (e.g. MRP and MRPII), with shop-floor data capture, and usually a more powerful computer to drive it all. This increasingly makes the organisation system-driven with decreasing scope for human interaction, particularly at shop-floor levels, when inevitable disturbances occur.

Thus if an incorrect schedule is derived by the system, the complexity associated with alterations to that schedule makes it almost impossible for the supervisors, operators, or even the planners to interact.

One alternative to this is to reorganise the factory layout on the basis of a product flow organisation with, ideally, products being manufactured complete in a single work island.

Sealectro Limited specialise in the manufacture of precision connectors and cable assemblies for communications signals at frequencies up to 46 GHz. Many of the connectors produced can be described generically as consisting of an outer body in which is assembled a central contact pin surrounded by a PTFE insulating bush. The demonstration island will be capable of manufacturing the contact pin/insulating bush sub-assemblies for a wide range of Sealectro's connectors, and, when transferred to Portsmouth and augmented by the inclusion of additional machines, tooling and other services, should be capable of manufacturing the full range of contact sub-assemblies.

The processes which will be represented within the island are shown below together with a list of the plant requirements for each.

1. Metal turning – Tornos Bechler Elector 16, Siemens CNC control
2. Metal Secondary Operations (e.g. slotting, milling, drilling and swaging) – Solma Combi-Matic, plug board sequential control
3. Heat treatment
4. PTFE turning – Tornos Bechler Elector 16
5. PTFE cross hole punching – Arbor Press
6. Assembly – Arbor Press
7. Epoxy resin injection – special purpose machine

Additional ancillary equipment within the island will include:

8. Wash and de-grease
9. Local storage facility – Kardex carousel unit – for storage of tooling, jigs and fixtures, finished products (i.e. contact sub-assemblies) and possible intermediate process storage where required (i.e. time buffering before bottleneck resource)

One process will have to be accessed by the island on a direct sub-contract basis under the authority of the island with a guaranteed turnaround time:

10. Plating

The demonstration island will be expected to make products (sub-assemblies of a saleable quality) within a specified and much reduced lead time, and the island will be capable of manufacturing approximately 100 different contact sub-assemblies. These sub-assemblies are used in a range of several thousand connectors with similar insulator/contact pin characteristics but with varying external features.

It is anticipated that the number of orders processed per week will be in the range of 50–100. The order quantities will probably be in the range 250–1000 parts. It is anticipated that only infrequently will more than one discrete order for the same product occur concurrently within the same time window and hence the typical batch sizes are likely to be in the same range as the order quantities.

Procedural Changes

The result of the simplification of product routings within the factory by reorganising the layout along group technology principles is that the overall control of the system is significantly simplified. Each island can be treated as a single resource at factory level with a consequent simplification of the overall factory planning task.

It is anticipated that the factory will operate in the following manner. Each island will be regarded as a quasi-autonomous factory-within-a-factory, with the island teams maintaining full authority and responsibility for all aspects of the island operations, including such matters as micro-scheduling, manufacturing engineering, and quality of all products produced.

Each island team will have service orders placed upon them by a single customer (the materials manager) who will provide a single interface with the sales department and hence the customer.

Orders will be distributed to the islands using the principle of repeated single cycle period batch control, i.e. only those parts which are required for assembly in the next time period will be manufactured in the current time period. This principle, however, dictates that:

1. Islands will complete all operations on any particular component, as scheduling different operations on the same component in the same time period in different islands could result in incompatible schedules.
2. The number of sequential islands in any product routing will be kept to a minimum as, under normal circumstances, the manufacturing lead time will be predetermined by the number of sequential islands multiplied by the length of the planning time period.
3. The minimum length of the planning time periods will be determined by the principle that it must be feasible for each island to complete its planned workload within the planned time period. Thus the planning time period can be no less than the longest batch process time including set-up time.

The materials manager will assemble orders for the next planning period from the sales order processing systems output. These orders will be exploded to island requirement level and will be allocated to the relevant islands on a rough-cut capacity basis. Once the order lists have been completed they will be passed to the islands for micro-scheduling purposes. The leaders of each island will be responsible for determining the optimum sequence of tasks to maximise the throughput of the island. A degree of interaction between the island leaders and the factory planner is anticipated after micro-scheduling has been completed to either overcome capacity shortfalls or utilise excess capacity in any particular island. Finally, therefore, each island will agree its workplan for the next time period with the factory planner, who will then be able to communicate firm delivery commitments to customers if required.

To undertake any particular job the island team will assemble its own manufacturing data list, consisting of design information, quality information, methods, and tooling information from either the island's own database or by direct access from corporate databases.

Each island will be provided with facilities to review, comment on and request changes to designs and will have access to a complex relational database which provides additional information on products beyond the simple geometric representation. This database with its associated classification system provides a common reference frame for both CAD designs and manually prepared designs and allows anyone in any island to easily access data on:

Parts – with graphics reference and change histories

Assemblies – with graphics references and change histories

Tooling methods – including tool details

Cultural Changes

Several important cultural changes are being identified, and an interactive education process involving the entire workforce is being developed. This process will involve the whole of the organisation, commencing at the top and including not only direct workers but also indirect workers such as quality, stores and toolroom personnel. Staff personnel are also being involved in this education process because it is considered that office functions could, and should, be structured on an island-type basis with work being managed in the same way as in the manufacturing islands.

The broad culture changes involved by moving towards a human-centred manufacturing organisation include the following:

1. Personnel will be encouraged to develop new skills in areas completely different from their traditional skill base. For example, operators will be encouraged to develop setting skills, such that the flexibility of manpower within each island will be maximised. The extent of this diversification of skills will, however, depend upon the desire and enthusiasm of each individual and, to a lesser extent, the encouragement of the island leader in providing different opportunities.

2. Each island, by virtue of being treated as a mini factory unit, will embrace all the functions of any small business including the following:
   Man management

> Work scheduling
> Elements of manufacturing engineering
> Routine maintenance
> Tool setting, maintenance and simple tool making
> Material management
> Quality assurance
> Elements of financial accounting

It is likely that several of these tasks will be carried out by personnel with other primary specialisations, e.g. the island leader tasks could well be carried out by the leading hand type of person.

3. Quality will be the responsibility of the island teams as a whole. Sealectro have already dispersed their inspection department throughout the manufacturing area as a precursor to this island autonomy.

4. Management responsibility and authority for output cost and quality will be devolved more to the shop-floor than at present.

5. Islands will be manned by multi-disciplinary teams of personnel drawn from several different traditional backgrounds. This will create problems for some individuals in terms of developing inter-personal skills, e.g. a skilled machinist from a traditional machine shop environment may have great difficulty in associating with assembly workers, the majority of whom are female. If an island team is to function as such, rather than as a collection of disparate individuals, then the development of such inter-personal skills will need to be encouraged with considerable sensitivity by the island leader.

6. Each island will be encouraged to manufacture products on a pseudo just-in-time basis using period batch control principles, such that, if insufficient work is required during a particular planning period to keep all resources (personnel and/or machines) in a particular island active throughout that period, then the excess time should be used for associated tasks such as preventative maintenance, set-up or process improvement, and manufacturing engineering rather than building excess inventory. Alternatively secondment of other staff may be considered.

7. The performance of each island team will assessed on the basis of its output quality and its due date performance.

8. Points 6 and 7 above make it impossible to use traditional individual financial inducements such as piece work bonus payments. An alternative inducement system is therefore needed to encourage the island team to determine a work schedule which maximises throughput or maximises the time for process improvement and maintenance rather than minimises the effort required of the individuals.

Technological Changes

Initial experiments in factory reorganisation at Sealectro Limited have revealed the need for technological change in that the shop-floor supervisors in pilot islands have expressed concern at the increase in workload caused by the devolving of responsibilities to the island team. This situation is almost inevitable if simplification of the manufacturing systems results in a greater diversity of tasks becoming the responsibility of the shop-floor personnel, and serves to illustrate the need for additional supporting technology to ensure that this increased workload does not eliminate any benefits gained from the simplification process.

It is anticipated that the supporting technology will include many of the facilities which will be available on the shop-floor monitor and controller.

The Rolls Royce Demonstration Site

R D Projects are developing a human-centred turning cell which will be demonstrated at Rolls Royce's Leavesden plant. Rolls Royce make aero and marine gas turbine engines, and the Leavesden division is a medium-sized semi-autonomous part of the major company and makes small gas turbine power units for helicopters.

The main cell components are two lathes, each with a gantry workhandler and carousel, and a cell controller. Each lathe will be fitted with a human-centred lathe controller developed at R D Projects. The two lathes, both of which will be fitted with live tooling and probes, are Matrix Churchill Two Series and Three Series machines.

The Two Series machine is fitted with an automatic tailstock, a workcatcher for use with bar work, a swarf conveyor and high pressure coolant for use with 'U'-drills. It has a twelve station, bi-directional turret (0.5 seconds face to face indexing) capable of driving live tooling toolholders on any station. Axial and radial live tooling toolholders will be available for drilling, reaming, tapping and milling. There is full servo control of the spindle 'C'-axis. A second six-station turret is also available for rear-end machining to components parted from the bar, and operations such as drilling, tapping, and boring can be performed. The Three Series machine is somewhat larger and has similar features except that it does not have a second turret.

Rolls Royce intend to use the human-centred turning cell for 50% of the production time on small batches (average 20) of precision components in sizes up to 200 mm in diameter. The remaining 50% of the time will be spent on development, preproduction, sub-contract work, demonstration and experimentation. The subcontract work will be aimed at a wide variety of component geometries in small batch sizes from one upwards; some of the sub-contract work could be prototyping of parts for new engines.

The cell will be set up in the main workshops and could be run in parallel with existing manufacturing facilities. Thus, human-centred and conventional turning can be compared on social, technical and economic grounds on the one site.

Social scientists from the Social and Applied Psychology Unit based at the University of Sheffield have made a contribution to the design of the human-centred lathe controllers, and will be helping to evaluate the cell at Rolls Royce.

The Human-centred Lathe Controller

R D Projects are developing lathe controllers which are being retrofitted to the Matrix Churchill lathes. Extensive graphical facilities on each lathe controller enable turners to carry out part-programming. Computer support is provided to remove routine programming work, but at all stages turners can override automatically-generated suggestions and set parameters based on their own knowledge and experience.

The geometry utilities allow the operator to define the geometry of the part to be made and the billet it is to be made from (Fig. 11.5). This geometry consists of a series of graphical shapes whose dimensions are defined by the operator. This is done by parameters, the values of which are set by the operator using pop-up tables. There can be multiple views of the geometries in different planes, i.e. end and side elevations to allow for the C-axis and live tooling. The order and type of the graphical shapes can be altered individually and their dimensions can be interrogated and changed. The geometry can be scaled and a zoom function creates a window so that a section of the geometry may be enlarged and smaller elements seen more clearly.

The drawing menu allows the user to create the following graphical shapes: line, arc, thread, groove, fillet radius and chamfer.

The CREATE program utilities allow the user to define the NC part-program via its component operations such as facing and roughing. These operations are defined by the user entering the values of parameters via the menu system. The system uses information from the user, geometric information from partshape, and tool information to calculate the tool path. When an operation has been defined, the resultant tool path is displayed graphically and the G codes are written to the NC part-program file.

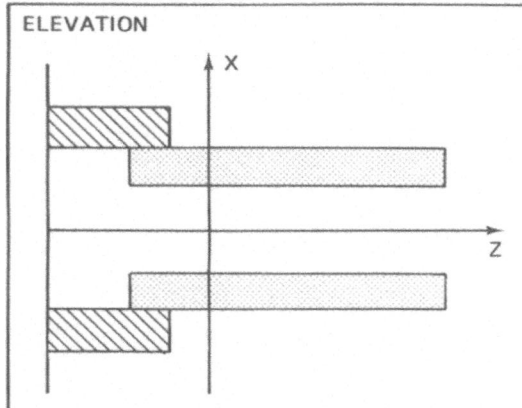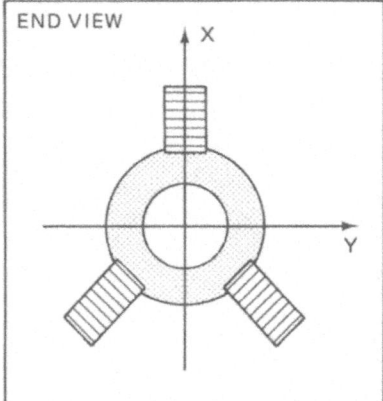

Fig. 11.5. A part geometry in two planes.

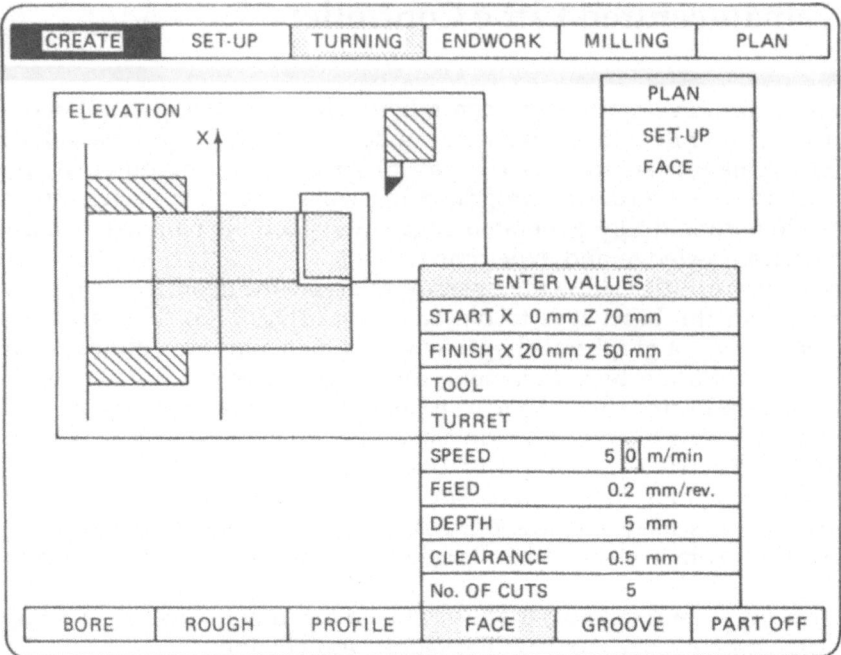

Fig. 11.6. Turning menu.

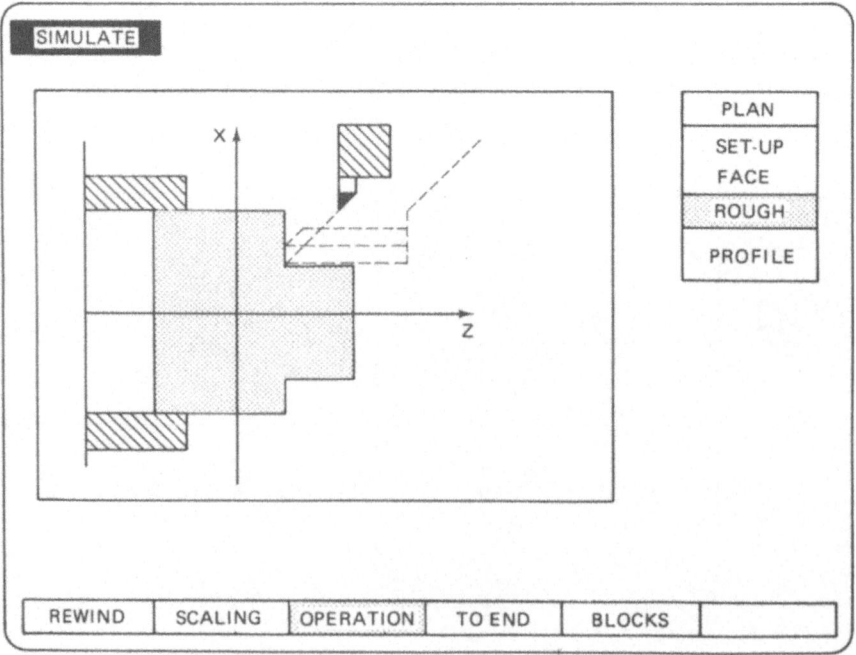

Fig. 11.7. Typical simulate display.

The turning and end-work menus allow the user to define the operations which make up the NC part-program (Fig. 11.6). For each operation the system performs several tasks. It selects a tool from the library and places it on a turret face. The plane in which the operation is to take place and its start and finish positions are chosen. The number of cuts, feed, speed and clearance allowances for the operation are then calculated. The user can change all these parameters. Once the parameters are accepted the system draws the graphical tool path and writes the G codes for the operation. The operations available are roughing, boring, profiling, grooving, centring, tapping, partoff, threading and drilling.

The system contains a tool library arranged as a table for each tool. Each table contains the tool's identification and geometrical information about its shape and orientation. Administration information shows the author of the tool table and the date it was created. A graphical representation of the tool is drawn from this information to aid the user in visualising the tool from its parameters.

A lifelike simulation of the blank, tools and fixtures is used to verify that the tool path will give the desired result (Fig. 11.7).

Concluding Remarks

This paper has presented an overview of a project which is developing CIM technology according to human-centred principles. Much work remains to be done in the last year of the project, and during this time there will be further theoretical and practical developments. At the completion of the project, it is planned that the demonstration sites will be evaluated against the project objectives using both commercial, and human-centred criteria.

Acknowledgements. Much of the material for this paper has been taken from internal documents produced by individuals working for organisations participating in ESPRIT project 1217 (1199) (see Appendix to this chapter) and their work is gratefully acknowledged.

References

1. Felix Rauner, Lauge Rasmussen and J. Martin Corbett, The social shaping of technology and work; Human Centred CIM Systems, AI & Society, Jan-Mar 1988, vol. 2, no. 1, p. 47.
2. Andrew Ainger and Shaun Murphy, Human skill and computer power, ESPRIT 87. Achievements and Impact, CEC, p. 1795.
3. F. Rauner, L. Rasmussen and J.M. Corbett, Crossing the border: the social and engineering design of computer integrated manufacturing systems (to be published in 1990 by Springer-Verlag).
4. J.M. Corbett, S. Ravden and C.W. Clegg, The development and implementation of human and organisational criteria in computer integrated manufacturing environments, in K. Rathmill and P. MacConnaill (editors), Computer integrated manufacturing, 1987, pp. 119–130 (IFS/Springer-Verlag).
5. Mike Cooley, The impact of computer aided design on designers and the design process, unpublished PhD thesis, 1979, North East London Polytechnic, London.

Appendix: Participating Organisations

BIBA – West Germany
BICC – Britain
Bremen University – West Germany
Greater London Enterprise – Britain
Krupp Atlas Elektronik – West Germany
NEH Consulting Engineers – Denmark
R D Projects – Britain
Rolls Royce – Britain
SAPU Sheffield University – Britain
Technical University of Denmark – Denmark
Teknologisk Institut – Denmark
UMIST – Britain

Glossary

C-axis	Axis of the main spindle, controlled in rotational position.
Electric discharge machining	A method of metal shaping in which metal is removed by spark erosion.
G code	Computer language devised for control of machine tools.
GHz	Gigahertz, a frequency of 1000 million hertz.
IGES	International graphical exchange standard – the format for data used in many CAD systems.
Live tooling	Driven tooling within a lathe used for off-centre operations such as drilling or milling.
MRP and MRPII	First and second versions of manufacturing resource planning, a widely used production planning system for planning at the factory level.
Pop-up table	Table of variable number of fields containing information, options, or blanks which is generated when a certain menu option is selected.
PTFE	Polytetrafluoroethylene – a tough polymer with stable properties to 350°C, frequently used as an electrical insulator.
'U'-drill	Specialist drill requiring high pressure liquid as coolant, and to assist with the removal of swarf.
2-D and 2.5-D	2-D is a two-dimensional sketch or model that contains no information about the third dimension. A 2.5-D model is the superimposition of several 2-D models, each representing one layer through a three-dimensional object.

Chapter 12

Postscript

Howard Rosenbrock

A number of significant achievements can be claimed for the project. First, it demonstrated that social scientists and technologists (engineers and computer scientists) can work together effectively at the design stage of a highly technical project with a strong research element. It is not unusual for social scientists to be associated with such projects, but it is very rare for them to be involved in the detailed decision-making process.

There are great difficulties in achieving this collaboration. The backgrounds of the two disciplines are so different that it is difficult at first to achieve a useful dialogue between them. Much time and effort, and much patience, are needed before the two sides can begin to appreciate and respect each other's point of view. The 'settling down' process by which this was accomplished persisted during the whole of the 26 months while the team was together, and was not completed at the end of this time. Nevertheless, in the latter stages the participants were working effectively as a joint social and technical design team.

In the process of 'settling down', the problems facing the two sides are naturally different. For the technologist, the problem is to include in the decision-making process of design a social dimension which is usually ignored. In some ways this is analogous to the inclusion of economic considerations, which forty years ago were often undervalued in relation to technical factors. Managements at that time constantly emphasised that if production costs and sales potential were ignored during the research and development and design stages, no later emphasis upon them could undo the damage.

This lesson was learnt, and is now a part of the accepted background in engineering. A similar integration needs to be achieved for the considerations which determine whether those working in the final system will be subordinate to the technical means (machines, computers, etc.) or will see them as tools which they can use to achieve their roles as a creative part of the production process. But the inclusion of this new dimension is more difficult than the earlier integration of economics.

Economic costs and opportunities are not exactly easy to quantify, but they are in principle quantifiable within the normal accounting procedures. The kind of benefit which is sought in the project, on the other hand, is not quantifiable in this way. It is easy to see (and Japanese experience has to some extent demonstrated) that an organisation which values and obtains and uses the initiative and ability of all those who work in it will be more effective, more flexible and more successful, other things being equal, than one which does not. Yet this benefit will never have appeared as an item in the accounts to justify the means which will achieve it. Moreover, the benefit which is sought cannot be made the only justification without an appearance of manipulation and exploitation which will probably prevent its achievement.

For these and allied reasons, the social considerations will tend to appear to the technologist as hazy and unquantified, and therefore difficult to balance against the technical and economic considerations with which he is familiar. For the social scientist, on the other hand, the converse problem arises. Generalities have to be made specific, and this has to be done in the face of very hard constraints.

The first constraint is time. In a rapidly developing technology, any drastic slowing down of the design and production and marketing of a product will make it uncompetitive. By the time it appears, newer designs using newer technology will have undercut its advantages, whatever these may be. Other constraints arise from economics and from the limits of technical feasibility.

Considerable freedom still remains, which can be used to produce designs which offer a more, or a less, satisfactory working situation. In exploiting this freedom, the social scientist may feel under pressure to make decisions which it would be more convenient to defer until the system is in use, and the opinions of workers can be more readily and reliably obtained. The fact that most of the available freedom is necessarily eliminated during the design process may be hard to accept.

These problems were compounded during the project by the fact that their magnitude had not been foreseen. Others subsequently following a similar path should be able to profit from the experience reported here. They can also draw encouragement from the fact that, in spite of the difficulties, it did prove possible for the team to work effectively. Something will be said later about the questions of methodology raised by the experience of the project.

The effective formation of a design team led to the second achievement which can be claimed for the project: namely that it produced a significantly better design of software for a programmable lathe controller. Details of this have been given in Chapter 4, and the following advantages are obtained.

1. Information from the CAD system is transmitted to the shop floor, not as code for machines but in geometric form. Conversion to code is performed at or near the machines, where information about their availability, suitability, and loading is most readily available.
2. When exceptions and difficulties occur, this arrangement allows a faster and more flexible response. Information arises on the shop floor and is made use of there.
3. In performing these functions, the worker is given full software support, but is left in control. Software support is retained to the fullest degree when the worker overrides any initial suggestions of the computer.

4. Though not implemented during the project, similar arrangements were envisaged for what was called 'microscheduling,' namely the determination of the most effective order in which parts (within the current job list) should be machined. Computer support under the control of the worker would again be provided.

Less actual implementation of software was achieved than had originally been hoped, for two reasons. First, the parallel application for funding by the Mechanical Engineering Department at UMIST (Chapter 1) was not successful. Secondly, the difficulties of integrating engineering and social science were greater than had been anticipated. Nevertheless, enough was accomplished to form the basis of the third achievement of the project.

This was the success in obtaining funds from the ESPRIT programme to support a further project (Chapter 11) which is intended to show a working demonstration, under industrial conditions, of a system designed according to the principles set out in the UMIST project. Without the experience gained at UMIST there would not have existed the confidence to put forward this application, nor if it had been put forward would there have been any hope of carrying conviction.

Though it cannot be called an achievement, one of the important results of the project was the identification of a number of problems which hinder progress towards the desired goal. One of them has already been mentioned, namely the problem of achieving a sufficient degree of agreement in outlook, in a design team containing technologists and social scientists, to allow design to proceed effectively. What is required is not unanimity, but only that extent of agreement which allows a fruitful tension between the two outlooks.

One of the stated aims of the project, in the application given as Appendix 2, was to develop a methodology for joint social and technical design. This was not achieved in the project, though an effective working relationship did evolve, and in retrospect the aim appears to have been wrongly stated. Even without the social dimension, engineering design is not amenable to codification as a 'methodology' – the best designers are often the least able to define what happens in the process of design.

What would now appear to be a much more useful aim to pursue is to find ways of shortening the 'settling down' process by which designers learn to work together in a new team. A process of this kind takes place even in a purely technical design team, but the addition of the social dimension makes it more difficult and more protracted. The facilitating and shortening of the process appears to be a subject deserving study by social scientists and by engineers.

A related question, more exclusively belonging in social science, is raised in Chapter 3: how can the experience in social science be expressed in a way which makes it easily applicable to the design process when this concerns a highly technical system? Where the object or system being designed is not highly technical, or where the technical options are confined to choosing one among a small number of known solutions, there is a wide experience in socio-technical design. The special difficulties of the UMIST project sprang from the highly technical nature of the subject and the long time scale. These factors made active participation by those, whether workers or social scientists, who were not deeply versed in the technology, both more difficult and less assured, because less well supported by their previous experience.

Finally, there were a number of questions raised by the project to which no answers could be found, nor any agreement about the direction in which answers should be sought. One of these was the 'hand and brain issue': that is, should the historical tendency to substitute mental for manual skills be resisted, or even reversed? Another was the 'blank table debate' described in previous chapters.

Both of these, and other similar questions which were not pursued to the same length, raised fundamental issues, and received different answers from different participants. It became clear that although all could agree in condemning certain features of our present technology, such as the subordination of people to machines, this did not lead to agreement on how a better technology might be defined.

Members of the Steering Committee and the Research Team came to the project with different backgrounds. There were differences in the field of specialisation, but also deeper differences in experience and philosophy and cultural background – strong individual influences of Christianity, or humanism or Marxism will be clearly discerned. In consequence, individual visions of a better kind of technology, and a better relation between technology and people, diverged markedly.

This diversity of view was a strength rather than a weakness of the project, in that what was achieved was probably more compatible with different future directions of advance than would otherwise have been the case. What was achieved represented a divergent path of technological development – not moving very far from the accepted path because of the limited time and resources available, but nevertheless beginning to define a different direction. This in turn was compatible with several further divergent paths conforming to the different views of participants.

Two comments can usefully be made on this, and in regard to future work. First, it is no doubt desirable that several different divergent paths should be pursued, at least for a time, so that comparisons can be made between them. Secondly, it is probably not very helpful to attempt to look too far ahead. If a truly divergent path of technological development is indeed opened up, it will be the views of those working with it, and living beside it in whatever capacity, that should carry most weight. These views will certainly change as technology changes, and with it people's expectations of technology. What is important is that at each moment the direction chosen is the one that they consider better rather than worse. In this continuing evolution there is always likely, as in the project, to be more agreement about the next step to be taken than about the ultimate aim.

Short CVs of Contributors

Allan Chatterton, after serving a technical apprenticeship, became a junior draughtsman and subsequently progressed to Section Leader. During 1967 he was responsible for the introduction of a new range of products. At this time, he became critically aware of the crucial importance of industrial leadership and motivation. Later his duties embraced personnel, industrial relations, and quality control. At the age of thirty-two years he became a works director and senior board member of a high voltage switchgear manufacturing company. Since 1975, in addition to his duties as a director, he has been researching into the causes, and cures, for Britain's industrial decline. These researches culminated in the award of a Masters degree from UMIST. In 1988 he was appointed Managing Director of his Manchester-based company. His qualifications include MSc, CEng, FIMechE, FBIM.

Mike Cooley was born in the West of Ireland and attended local schools before his higher education on the continent. He studied and worked in Switzerland, Germany and England, qualifying as an engineer, and subsequently whilst working in industry obtained his PhD. For some eighteen years he held senior design and R&D posts in the Aerospace Industry and has a number of patents for products and systems that he designed.

Until recently, he was Director of Technology at the Greater London Enterprise Board and Chief Executive of TECHNET Limited. He is currently a director of several high-tech companies, and on a voluntary basis of the Technology Exchange, the Third World Information Network, etc. He writes for a wide variety of publications and broadcasts on radio and TV in several countries, particularly West Germany. He has collaborated on an *Equinox* film on Brunelleschi and the nature of skill. He has produced over 50 scientific papers and is author or joint author of eleven books on technological change in German and in English. His work has been translated into over twenty languages.

He has been a guest professor and held seminars at universities throughout Europe, the United States, Australia and more recently, Japan. He is currently a visiting professor at UMIST and guest professor at the University of Bremen. He is an advisor and consultant to several companies and Governments. He has always been concerned about the impact of technology on society and the environment. He has been an active trade unionist, and was the lay member National President of the Designers' Union. He was one of the primary authors of the Lucas Workers' Plan for Socially Useful Production. He was one of the instigators of ESPRIT project 1217 (1199) – Human-centred CIM.

Amongst his awards and distinctions are the $50,000 Alternative Nobel Prize in 1981 (which he donated for socially useful production) and the Keys of the City of Osaka in 1987.

Martin Corbett is an industrial psychologist who studied at the Universities of Leeds, Lancaster and Bath before joining the UMIST project team in 1982. Since leaving UMIST he has completed a three year research fellowship with the Medical Research Council at Sheffield University examining organisational and psychological aspects of computer-aided manufacturing. He is a member of the IFAC Technical Committee on Social Effects of Automation and is currently Lecturer in Organisational Behaviour at the University of Warwick Business School where he is continuing his research work on human-centred technology.

Håkon Finne was born in 1952 and is Norwegian. He was educated in Applied Physics and Production Engineering at the Norwegian Institute of Technology (NTH), Trondheim, and also took a number of Social Science courses at the University of Trondheim. He worked as a research assistant in the Division of Psychology and Social Science at NTH, 1974–1975, and subsequently as a school teacher, programmer and laboratory worker. He also worked as a research scientist in the Institute of Social Research in Industry (IFIM) in Trondheim. In 1981–1984 he held Research Fellowships, working in Manchester, and at NTH. From 1985 he has been a research scientist at IFIM, and has spent time at MIT. His interests are in new technology, quality of working life, worker participation and organisation development, with basic research in the social construction of technology. His publications include co-authorship of three books (in Norwegian) on social science for engineering students and technology designers.

Roger Holden took his first degree in Electrical and Electronic Engineering at the Polytechnic of the South Bank. Subsequently he gained a Masters degree in Computer Science at the University of Manchester before returning to the Polytechnic of the South Bank to work for a PhD in collaboration with Cableform Limited. This was followed by three years as a research assistant at UMIST working on the project described in this book. He is now employed as Senior Programmer by the Wythenshawe Division of Ferranti Computer Systems Limited, were he is involved in the design of process control systems. Outside work, he is interested in industrial history, having published two papers in this field and holding a Certificate in Local History from the Department of Extra-Mural Studies, University of Manchester. He is a keen walker and attends Stockport Meeting of the Religious Society of Friends (Quakers).

Paul Kidd was from 1970 until 1975 an apprentice electrician and undertook five years of part-time study on a City and Guilds Electrical Technicians course. On completing this apprenticeship he read for a Bachelors' degree in Instrumentation and Systems Engineering at Teesside Polytechnic, and graduated in 1979 with first class honours. As part of his undergraduate training he spent about eighteen months working in the Nuclear Power and Chemical Industries. In 1979 he went to the Control Systems Centre at UMIST to read for a Masters' degree in the Theory and Practice of Automatic Control. He spent two more years undertaking research into the robustness aspects of control system design, for which he obtained a PhD in 1983. From 1980 until 1983 he was also employed as a research assistant in the Control Systems Centre, working on the development of multivariable control systems for the MOD. From 1983 until 1985 he worked on the UMIST research project with which this book is concerned. In late 1985 he took up an appointment as Lecturer in the Department of Control Engineering at the University of Sheffield. He now works as an independent researcher and consultant. His research interests are in the area of engineering and information technology, with particular emphasis on the use of social criteria in design, and the development of social and technical design methodologies.

Lisl Klein is a Senior Social Scientist in the Tavistock Institute of Human Relations (TIHR). Before joining the Institute she alternated between work in industry and in academic settings. In industry she has spent a year on the shop floor, three years as a factory personnel officer and five years as Social Sciences Adviser to Esso Petroleum (UM).

Research has been on the human implications of work study, on the behavioural consequences of management control systems, and on the utilisation of social science. Her work since joining TIHR has included eight years consulting to one of the UK clearing banks, six years as a consultant to the German Government's programme to Humanise Life at Work, two projects on job design in the building of new plant, consultancy on organisation to a research centre and to the Alvey programme.

Books: *Multiproducts limited* (HMSO, 1964); *New forms of work organisation* (Cambridge University Press, 1976; also in German); *A social scientist in industry* (Gower Press, 1976; also in German); *Social science in practice* (with K.D. Eason, Cambridge University Press, to be published 1989).

Shaun Murphy has a BSc in Chemistry with Social Studies from Queen Mary College, London, and a PhD in Theoretical Chemistry from Cambridge University. After leaving college he was responsible for advising on the use of computers and automated equipment in a large scientific organisation, and for the last five years has worked for Greater London Enterprise where he has had responsibility for a number of large technology-based projects, and investments in technology-based companies. He is currently Project Manager for ESPRIT project 1217 (1199).

Harold Palmer was born 24 July 1924. Qualifications GRSC 1945, FRSC 1951, C Chem. Entered manufacturing industry in 1939–1981, firstly, as a laboratory assistant and latterly as manager of technical innovations. In 1981–1984 became coordinator to the SERC-ESRC Joint Committee. His main managerial appointments are:

1950 Section Manager Research Department, British Nylon Spinners
1958 Associate Research Manager, British Nylon Spinners
1962 Pioneering Manager, British Nylon Spinners
1967 Technical Planning Manager, ICI Fibres
1972 Manager R&T Strategy Unit, ICI HQ.

Howard Rosenbrock was born in 1920 and educated as an Electrical Engineer at University College London. After service in the RAFVR from 1941 to 1946, he worked in industry until 1962, being Research Manager for CJB from 1957. In 1962, he joined the Control group under John Coales at Cambridge, where he remained until 1966, with a year at MIT in 1963–1964. From 1966 until his retirement in 1987 he was Professor of Control Engineering at UMIST, where he set up the Control Systems Centre. He is author or co-author of four books and about 130 papers. Qualifications include DSc 1963, FIEE, FIChemE, FInstMC, FRS 1976, FEng 1984. With Mike Cooley, he assisted in formulating the ESPRIT project.

SERC Application

Project Title

A Flexible Manufacturing System in which Operators are not
Subordinate to Machines

Description and Objectives

To develop software which will enable the operator to program a flexible
manufacturing system by making the first of a batch of parts. In doing so, to
develop a methodology for the simultaneous consideration of social and
technical aspects during the development of new technology.

Case for Support

The Development of New Technology

This application is based upon the proposition that new technology can be
developed in two different ways. (By 'new technology' is meant machinery
incorporating new developments in hardware or software. By 'machinery' is
meant any mechanical or electronic device: for example an FMS, a CAD
system, an assembly system using robots, etc.). In the first way, those human
beings who will work with the machine are made subordinate to it. In the
second, the machine is made subordinate to the human.

The first way is the one that is almost universally followed at present.
Human subordination is sought in a number of ways. Work is paced by the

machine. It is subdivided and rigidly specified so that it takes on a machine-like character, complementary to the operation of the machine. Situations requiring the human qualities of skill and judgement are eliminated by standardisation wherever possible, and where this cannot be done they are removed from the place of work and the necessary decisions taken elsewhere.

Examples of the second way hardly exist at present, though they existed at the beginning of the Industrial Revolution (Hargreaves's spinning jenny, Crompton's mule). The human qualities of skill and judgement are not eliminated, but are assisted and made more productive. They are not removed from the place of work. It is accepted that total standardisation is not achievable (machines, for example, may be out of action, or parts required may not be available in the specified order). Human intervention to deal with these abnormalities is assisted by any means which will make it more effective (for example, computers to assist in the re-scheduling of work).

Some early research in computer-aided design was aimed in the second direction. Some work in Artificial Intelligence (for example, computer-aided diagnosis in medicine) adopts this course at present. In both cases, the choice of the course to be followed has not been fully explicit. It has stemmed from the research worker's implicit assumption that he was designing a machine for himself to use (just as Hargreaves and Crompton did). Whether subsequent development in these areas will go in the second direction rather than the first seems highly doubtful, in the absence of effort directed explicitly to that aim.

The aim of the research proposed here is to develop a flexible manufacturing system in which the machine is subordinate to the operator. As research aimed explicitly in this direction hardly exists, a major task will be to develop a methodology for research having such aims.

Engineering Research

The view suggested is that engineering research is a malleable process, which can be diverted to follow different paths. It is in essence a problem-solving procedure, which starts with an aim (for example, develop an FMS cell) and a tentative route for achieving it. In attempting to follow the proposed route, various difficulties will be uncovered. These will seldom be absolute bars to progress in the original direction, but they will make some deviation seem more attractive than the straightforward path. The decision in each situation of this kind is made in the light of the general aims, and the means available for overcoming the difficulty. As a result of this process a feasible path is gradually uncovered, perhaps with some back-tracking but not much. The path that is followed, and the ultimate solution, therefore depend on the difficulties that are met, the criteria by which they are judged, and the technical means available.

The important point is that the research does not proceed by evaluating all choices at the first difficulty, then all at the next to be met for each of the first choices, and so on, and finally choosing the best out of all this multitude of branching decisions. As in chess, the exponential growth of choices makes this infeasible.

Two consequences follow. The first is that the socio-technical approach, as usually described, cannot be applied. That is to say, it is not possible to list the technological possibilities, and then match these with different job designs, and finally choose the best combination. This is appropriate, for example, where the technological choice consists in selecting from among commercially-available machines. It does not apply in research, where in reaching a feasible solution, a multitude of decisions have already been made which have implications for the relation between machine and worker.

Secondly, the path followed, and the solution reached, are sensitive to the criteria used at each of the branching choices. If social considerations are introduced into these choices (that is, considerations of the relation between machine and worker) then a different path will be followed and a different solution reached.

This may not be as obvious as it should be, because social considerations are not usually included at the research stage. The analogy with economic considerations will make the situation clearer. It could be argued that engineers should concern themselves in their research only with techno-logical considerations, and should present the technological alternatives for a final economic evaluation. It is well known that this will not lead to the desired result: economic opportunities will have been neglected along the way. Much management effort is given to ensuring that the detailed course of the research is governed by economic considerations. What is suggested in the proposal it that it should be governed also by social considerations.

What is proposed is that in an engineering research project to design a flexible manufacturing system, the successive decisions needed to overcome the difficulties which arise can be taken not only in the light of technical and economic requirements, but also of social requirements (as defined above). Success in this project would consist in producing an FMS which satisfied the following criteria.

(i) It was technologically satisfactory in performance, reliability, etc.
(ii) It was economically satisfactory and would be attractive for use in a small engineering shop.
(iii) It did not subordinate workers to itself, but instead acted as an aid to them in making their skill and ability more productive.

Given success in this sense, a number of research topics in social science could be posed. For example, what sort of people select themselves into this kind of work, and what kinds of adaptation and adjustment do they make in their work? These would be interesting, but would probably have to wait until the machine was in industrial use. The critical decision would therefore come earlier: knowing by demonstration that the proposed aims were feasible, would society choose to pursue that direction? 'Society choosing' would probably be expressed in a number of ways: adoption by some employers, pressure from some unions, a change of viewpoint by some engineers, and so on.

A background difficulty for the above proposal is that a number of arguments are put forward, from economics, or politics, or other areas, to suggest that the aims are impossible to meet. These arguments are briefly considered and briefly answered in the Appendix.

The FMS

The FMS upon which the proposed research would be done already exists in the Mechanical Engineering Department at UMIST, and is the subject of a grant by SERC to Professor John Davies. It consists of an NC lathe, NC milling machine, Unimate robot, and PDP 11–34 computer.

The NC lathe has been and the NC milling machine is being, designed so that it can be programmed by the operator, who does this by making the first part in a batch. The computer records and plays back the operations as necessary. This has operational advantages as compared to the alternatives, in which programming is done by specialist part programmers, or as an adjunct of CAD. These alternatives require the program to be verified and possibly amended and verified again: programming by record/playback includes immediate verification. Also, when a number of NC machines of different types are available, the operator on the shop floor has better knowledge of which one should be used than a parts programmer or CAD operator. Machine utilisation can also be improved with record/playback, because in the alternative systems a mismatch can occur between programs and parts for machining. With record/playback, if this occurs, the operator can at once proceed with the making of the next part.

These advantages all spring essentially from the fact that knowledge which is needed for programming the FMS largely arises on the shop floor, and is best made use of there. Professor Davies, and the group working with him, have already been led by these considerations, and by a concern for the operator's job, to design the hardware and software of the lathe and milling machine in such a way that the operator's knowledge and skill can be effectively used. They propose also to follow a similar path in the design of the FMS and they have the hardware and software skills for this purpose. What is proposed in this application is that there should be added, to the existing research, the professional ability of a social scientist in considering the relation between man and machine, and the further professional ability of computer specialists in evolving suitable software.

In the fully-developed system, it is envisaged that information about the part to be made would be obtained by the FMS operator from a drawing, or later from a display connected to the CAD system. He would program the manufacture of the part by making the first of a batch, using the NC machines as described above. Programming of the robot would be done at the same time, probably in part by instructions entered through the computer, and in part by physical lead-through. The second would alleviate the requirement on the robot for accuracy if it is programmed off-line: final adjustment of the robot position in relation to the machines could be made manually. The FMS operator would probably also be given some responsibility for re-scheduling production in the light of availability of machines, parts and tools. He would be given computer-assistance for this purpose.

In reacting with NC machines and the robot, the operator would to a large extent communicate through a computer interface. This would give great flexibility, so that the system could be changed rapidly and easily. The operator could be offered a choice of modes of working. Some of the techniques of AI could probably be used to make communication between operator and system more easy and natural.

Such a system would, it is believed, make the machine subservient to the operator. He would develop a skill in using the machine, which would aid him and make his work more productive. At least, it would be easy to use the system in this way, though no doubt it could also be used in a way that did not have these characteristics, given a will to do so.

Methodology of Research

What has just been described corresponds to the aim of the engineering research, defined as it was earlier. In pursuing this aim, difficulties would continually be faced, and would have to be overcome. It is proposed that in tackling such difficulties, regard should be paid to technological feasibility, economics, and also to the aim of subordinating the machine to the operator. The first two are standard, but the third is new, and a methodology appropriate to it does not exist. It is in this area that lie the largest uncertainties of the research. By attacking this problem for the particular example of the FMS system it is hoped that the groundwork for a general methodology can be laid.

The difficulty lies in predicting how a given decision (perhaps on a matter of detail) will affect the relationship which the operator perceives between himself and the machine. For example, the computer system should protect the operator against damaging and costly mistakes, yet should not interfere with his freedom of action, and these requirements are partly in opposition. Ideally, one would like to achieve dialogues such as the following:

comp: I am predicting a collision between the tool and workpiece. Are you sure the next move is safe?

op: Why collision?

comp: Envelope of workpiece intersects envelope of tool at 35% motion of robot.

op: Show me
        (graphic display)

op: Envelopes are conservative. Move is safe.

comp: OK, but I suggest we run it slowly first time, with your finger on the stop button.

op: OK. Go.

This degree of 'intelligence' in the computer is probably not achievable at present, and a compromise would have to be sought, while preserving as much as possible of the relationship suggested. Similar problems arise in the socio-technical design of systems, of which the above is clearly an extended version. It would be the task of the social scientist when such problems arose to add, to the technical and economic considerations of engineers and software specialists, knowledge from the social sciences on the human consequences of different possibilities. He would also assess, so far as it could be predicted, the effectiveness of the resulting system in social terms. He would moderate the activities of the technologists and interact with them, keeping before them such considerations as:

do the jobs which people will be doing in the new system provide opportunities to cooperate with others?

do they provide opportunities to learn new things, and develop?

do they provide feedback about one's own efforts and the performance of the whole system?

do they provide a range of experiences or only repeated, limited experiences?

do they permit control over such things as pace and movement?

do they provide an opportunity to use judgement and make decisions?

In attempting to answer such questions, the socio-technical approach favours the direct involvement of workers in the design of the system. A difficulty in the situation envisaged here is that the appropriate group of workers does not exist at the research stage. Possible solutions will be sought by the following and probably also other means

(i) By using the experience of proxy workers (which can be arranged by Allan Chatterton, below).

(ii) By drawing upon the experience of workers where analogous situations to those being studied already exist in industry. This could be arranged, through Trade Unions and otherwise, by Mike Cooley.

(iii) By a systematic investigation of alternatives, using simulation where this is possible; for example in computer interfaces.

(iv) By leaving the greatest possible freedom for subsequent changes in the system as it reacts with the operator. The existence of computer interfaces with the operator makes this easier than if the interfaces were all through mechanical hardware.

Because the scope of the research is very wide, it is proposed to set up a steering committee to guide it. The following have agreed to serve on such a committee, and one or two further members might be brought in subsequently.

Allan Chatterton, Works Director, Long and Crawford, Gorton Road, Manchester.

Mike Cooley, Lucas Aerospace Shop Stewards Committee.

(John Davies, Professor of Mechanical Engineering, UMIST).

John Fox, Research Computer Unit, Imperial Cancer Research Fund, London.

Lisl Klein, Tavistock Institute of Human Relations.

(Howard Rosenbrock, SRC Senior Fellow).

Two preliminary meetings have been held, and this research proposal has been agreed by the committee members. It is envisaged that when the work is completed, members of the committee would contribute accounts of the work from their different viewpoints in order to give a comprehensive record. The proposal is seen very much as a first step: ultimate and long-term success would be shown by the routine consideration in all engineering research of the issues raised here.

This proposal relates to the Senior Fellowship awarded to Howard Rosenbrock from January 1st 1979, and the research staff who are requested are those mentioned in the Fellowship application. Because the present application is so heavily dependent on the earlier proposal by John Davies, it is signed equally by him.

Three research assistants have been requested starting at point 5 on the RA [1A] scale. Two of these would be engineers with computing experience, or computer scientists, either postdoctoral or with about 5 years industrial experience. Both should have an interest in the social aspects of their work. The third would specialise in the social aspects, and could come from a variety of backgrounds. A social scientist with industrial experience or an engineer with subsequent training in social science would both be suitable. So would other types of background: this post would be advertised in wide terms.

Visits by the research workers to sites in Europe where similar work was in progress would be highly desirable, and have been allowed for. Four meetings each year of the steering committee have been allowed for. Some extra materials will be needed for tests, particularly if proxy workers are used, and £2000 has been allowed. Other requirements for equipment etc. are already provided by Grant No. GR/B11585 to Professor Davies.

Appendix – Objections Which May be Raised

A number of arguments exist which purport to show that what is proposed is impossible. The more important may be summarised as follows.

(i) Technological determinism. There is only one technologically feasible route to follow. Alternatively, if there is more than one route, one is technologically superior to any of the others.

(ii) Economic determinism. Technology develops through competition in the market-place. The successful development is therefore the one which is economically most advantageous. Any alternative will be economically inferior.

(iii) Political determinism. The choice of the first type of technology rather then the second stems from the self-interest of those making the decision. Only when this self-interest is eliminated will the second choice become possible.

(iv) The characteristics of the first type of technology spring from the authoritarian attitudes of engineers. Any alternative proposed by engineers will be equally authoritarian. Only by participation of the workers concerned can a valid alternative be developed.

The telegraphic brevity of these summaries is not intended to conceal the force and subtlety with which the objections can be put. It will also be recognised that various interest groups are associated with asserting or denying any of these propositions. The answers to the objections which will be suggested here are, with equal brevity, as follows.

(a) The nature of engineering research is considered in the application. If the picture given there is correct, then technological determinism is false.

(b) The economic incentives acting upon an enterprise do not necessarily lead to the fullest use of the national resource which is represented by the abilities of its people. Education provided by the State, for example, contributes to national prosperity without necessarily affecting the relative competitive advantages of enterprises within the State, so that none of these individually has an incentive to use its own money for education. Similarly, if all enterprises used a technology which made better use of human abilities, their relative competitive advantages might be unchanged, but all might be more prosperous.

(c) No better way is known of balancing different interest groups than the one that has evolved in Western Europe, imperfect as that may be. If something is strongly desired by enough people, it can probably be achieved.

(d) If offered the choice between two kinds of technology, society will choose one or the other. At present the existence of a choice is not apparent, and it can only be made apparent by technologists.

It is unusual to raise such wide issues in an application, but no apology is made for doing so. What is proposed obviously has wide repercussions, and the foregoing considerations would undoubtedly be in the minds of those reading the application. It seems best, therefore, to face them openly.

Transcript of a Part of the Steering Committee Meeting Held on 12 July 1982

What follows is a verbatim transcript of a part of the Steering Committee meeting held on 12 July 1982. It is included because the meeting decided one of the major questions raised by the project, namely the way in which a contribution from workers should be sought and incorporated in the design of the system.

Two difficulties arose in preparing the transcript. First, some speakers (notably HF and JF) were recorded very faintly, and it has not been possible to present their contributions fairly. Secondly, spoken English is very different from written English, and as with all verbatim transcripts, some of the information is lost from the record. The spoken words are accompanied by gestures, facial expressions, differences of tone and of pace which are an essential contribution to the meaning but are lost in the transcript. It has also to be remembered that participants were not speaking from prepared and fixed positions, which could be accurately defined, but were engaged in a process of mutual accommodation to reach an acceptable conclusion.

It is hoped, nevertheless, that what is given will help in understanding how and why decisions were arrived at. The following were present:

JB   John Boon
AC   Allan Chatterton
MC   Mike Cooley
JMC   Martin Corbett
HF   Håkon Finne (observer, by invitation)
JF   John Fox
KG   Ken Grainger (part-time, by invitation)
LK   Lisl Klein
HP   Harold Palmer (Chairman from 4 p.m.)
HR   Howard Rosenbrock (Chairman)

HR Well, I suggest that we close the session on this, otherwise Ken Grainger is not going to get the opportunity to say what he wants to say to us. And so I'd like to ask him – he did come and he spent an afternoon with us, and made a point very strongly that perhaps we were not taking enough direct input from the shop floor so to speak. Now I will leave it to you entirely, Ken, as to how you want to introduce this.

KG Well, I came here as a volunteer from Coventry Workshop, which is an organisation funded by charities, providing research and educational facilities to shop stewards' groups and tenants' groups in Coventry. Now I'm working with a group of stewards in the machine tool industry, that's part of my brief with the Workshop, and part of the reason why I came up to ask about the UMIST project. I was interested in the UMIST project because of the whole idea of operator programmable lathes, and developing that with assistance. So the questions I had in mind were about the social scientists, I was very curious about the role of the social scientists in the project. I wasn't quite sure what their role was supposed to be and the question that I also had, was why weren't machinists actually involved in the design of the system. So really it was that main question that I wanted to pose again, and take up the discussion from there. I asked the question because I was actually struck by the political perspective of some of the people on the project, and it seemed to me that an important aspect of the project, what made it perhaps unique in this sort of research, is that it offered something to Trade Unionists to develop as a positive proposal when they are in collective bargaining over technical change. So it seemed to me a key aspect is really about what is the nature of machinists' involvement, of workers' involvement in this project. And I could think, at the time I was talking to Howard about the project, I could think of plenty of machinists in Coventry, in the machine tool industry, who would have the experience of actually working with manual data input machine tools, of working with designers on prototype work, who could hold their own certainly in discussions about designs of these systems.

HR Your suggestion, as I took it at the time was that we ought to have associated with the project either as a member of the Steering Committee, or more directly working with the RAs, somebody who was, who had direct shopfloor experience.

KG [Assents.]

HR The difficulty I had with that was that, you know, from the beginning I had felt that somebody in that situation would be, tend to be, out of his depth in terms of the computer technology, the development of the system, but maybe I am wrong about that. And it turns, it seems to me to turn, on whether one can take someone with that kind of experience and expect, hope that he will be able to make a direct contribution to the project. It would seem to me that it would be easier at the Steering Committee level because in a sense we're all laymen with respect to other people's areas. With respect to Lisl's area for example, or the AI; and other people, with respect to other areas. So someone who has only a partial background is only in the same situation as everyone else. When it comes to a closely-knit working team – we advertised at the postdoctoral level – that would seem to be a much more difficult project. Do you, do you have that, you know, would you agree with that, or do you feel that's . . .?

KG Well, I don't really feel that the kind of people I am thinking about would be out of their depth on the design stage.

JB I rather like the idea of this project diverging to the problem Mike and I have, we can have, imagine several levels in which the machine itself is in control, sounds the right sort of area to go to, if you like, from our point of view, say well here let's see what they think about handling, controlling, manual, in play-back mode, let's look at *our* system on the machine and get their reactions about it. Let's look at an MDI sort of approach. And I am rather attracted to that as one of the ways in which we can get a reaction to the views that Mike and I have and if you like, also influence the design. Now, if you like, what I am trying to do is not put it into the project, I am trying to get some sort of knowledge base, on John's lines. Let's get some practical knowledge how people might react to these various systems. And that might fit . . .

LK I've got two worries. Sorry. I've two worries. One is about tokenism, and I think we must be very careful not to do . . .

? Cobblers!

LK . . . this, you know you have got to have a black on the board, and you've got to have a woman on the board, and who can argue against that?

HR A black woman speaking Spanish is the ideal [laughter].

LK You know, I mean what kind of a rat are you if you raise an argument against it, because it's an ideological point. So yes by all means, if it's for real, but not just because it is a difficult thing to argue against. Now where it is for real, in the instance you are talking about I would actually be worried because you would have your one experienced, splendid, interested, etc. machinist whom you are going to look at as your representative machinist. And he's not, he's a bloke.

JB No, I was looking at several machinists . . .

LK And you do need at that stage to find a way of getting the reactions of, and the kind of choices they would make, but you don't do that by putting your one machinist on the team.

JB Yes, alright, I'm sorry . . .

LK You're going to get one prejudiced individual like they're all prejudiced.

JB I'd rather adapted Ken's argument, to sort of putting, if you like, a practical machinist onto the team by saying let's go and see Ken's machinists.

LK Oh fine, lovely, that's fine.

JB Yes, I . . .

HR   But that's something rather different, though, isn't it?

LK   That's quite different. He's saying put one of them on the team.

JB   What I do accept is we need that experience, Ken's machinists, so I wouldn't, I see myself sort of having to fit that mould, the machinists' mould, by saying this is what we do, feels it should do, can do. And I'd much rather have a machinist tell me in particular what he thinks it can do. I would rather go to the machinists I think, and I think there's a lot of this sort of problem, of do we or do we not have artificial intelligence types of the questioning and answering, because a machinist won't really know that that's what we are talking about. Some of the things that we want to sort of perhaps put into design or not: but not specifically machine orientated, something that's within his experience. Now I know that they're widely based individuals but we are going to sort of stretch them a long way in development outside their field of experience.

KG   I'm not sure ...

JB   You obviously are not.

KG   I am not sure that's true because I mean in some of the machine tool plants where they've been developing MDI Systems, people have got experience of previously working on manual, then working on CNC systems. Systems which have been designed deliberately to cut the operator out, and then they have seen a complete reversal [word missed] with MDI where the operator is supposed to come back and the programmer is cut out. So there are some people around who've got experience of those sort of technical changes being put through on the products they are making. And I would have thought ...

JB   Yes, the [word missed] that we've got on the medicals' example and efforts. We might have a doctor saying, asking the computer what's wrong with this particular patient, or we might change, turn the coin right over saying here's the patient now asking the questions, and it's not the machinist anymore, it's sort of doctor and patient or machinist and some-body who's not a machinist at all. That's the problem I think we could be throwing the machinists who are actually in the Steering Committee [voice fades].

KG   I'm not quite sure what the parallel is.

JB   Something that's not actually related to the machine tools at all. The use of, how we're going to apply the computers to the system, [voice fades], is it going to be a Cambridge ring or what the hell are we going to do to get these things to talk to each other. We probably are going to leave the machinists a bit cold, probably going to leave Martin a bit cold, I don't know.

KG   But it would be important to actually translate those terms back into points that Trade Unionists could talk about because if an important aspect of the project is actually relating it to Trade Unionists as part of something they would carry on in collective bargaining, then surely the closer the links between the project and Trade Unionists the better.

LK   Ah, that's an entirely different point. If the Trade Union wants to keep an eye on us, and be linked, that's quite different from having one individual machinist.

KG   Well, I would have thought that that individual machinist wouldn't be an individual machinist simply, you know, parachuted in, but I would think it's important to have links with a Trade Union group, and the rest . . .

LK   Oh, that's . . .

HR   That's a different group, that's a different point, I think. Yes, and it is something we have talked about and, you know, we are proposing to make a presentation to a number of Trade Unionists later on. But the point which stuck in my memory after we talked was that I had looked at the matter in this way; that we were setting out to try to design a system which was not only economic and technologically satisfactory, but also provided a satisfactory kind of work for the user, the operator, and didn't constrain him by subjecting him to the machine, allowed him to develop, allowed his skill to develop and so on. Now, it's, I had looked on the matter this way, that in making decisions and choices with respect to the development of the technology, we had to take those considerations into account and that was a question within social science. Now, at the time I think you rather questioned that, and said it's not so much a question for social science, it's a question for the people concerned, if I wasn't misinterpreting you. But then the difficulty arises, can the people concerned directly take part in the discussion?

LK   What was your question about social science, it was the first point you made, was it . . .?

KG   Well, I'm somewhat worried about the idea of social scientists actually presenting an objective position, a sort of a neutral, politically neutral position when they are evaluating the response of machinists to the new systems, and I was wondering in a way whether social scientists were presenting themselves as proxy workers, in that sense, of you know reinterpreting machinists' response to a system to designers, saying well, you know, the machinists really feel this and this about the system.

LK   Aren't you jumping to conclusions [word missed]? Is that what you hear us saying?

JF   Yes I wonder if [word missed] we may have projected an incorrect image. It's all very well to use words like this. At the moment I don't think we're doing design. It may be we'll be doing design in two years' time, when we know what the design options are. At the moment we don't even know that. We're not in a stage where somebody in a company is saying OK we'll have to take a bunch of well-known techniques, put them together and make a new product, and then people have inputs and say what sort of product they want, because at this stage we don't even know what the options are, there are just so many problems, it really is a jungle, to use the word that Howard carefully avoided this morning, [laughter] I mean, it really is a jungle. And to bring somebody in . . . One of the things, everybody around this table in one way or another has in common is, we are all in one way or another trying to do research, and that sort of, the uncertainty that comes with that, is one of the things that I think we can cope with. It seems to me that when we get to design where we are actually putting forward solutions rather than trying to agree on the questions, then that might be an appropriate point at which to find some method, not necessarily by inviting someone to join the group, but some way of going out and saying to people 'look, these are the options, do you have any comments?'. But actually we need to circumscribe the possibilities beforehand, and I think many people will get impatient, who haven't got a research training, a research background, and wish to not spend time discussing the finer points of methodology.

HR   Håkon?

HF   I think what John is asking for now is for workers and unions to give their reactions, rather than giving their actions [words missed], which gives us a sort of an asymmetrical relationship with these people, in the sense that first the esoteric activity of the research goes on and you come out with a set of options left that someone else can pick on.

LK   Oh, wait a minute, wait a minute. I've had a lot of experience of trying to get an active contribution from, and have always met the response, give us something to react to, give us a choice, and we will tell you what is preferable, but it is very difficult for us even when the will is good – and it isn't always – it is very difficult for us to be pro-active. Secondly there is a stereotype about social scientists, partly among some journalists and partly among some professional competitors like industrial relations specialists, which sets them up as being proxy workers. Now that is not the contribution of social scientists. I am not aware of any attempt to set up a proxy worker. The contribution is a methodological one, and the feeding in of research experience and what little design experience there is, it's not saying workers think this, it is saying if you want to harness the experience of workers, give them options and see what they do, don't make guesses about it, which is the sort of discussion we've had here. That's where scientific methodology comes in.

AC   Could I suggest a compromise. I feel, being a little provocative, I feel that Ken's presence this afternoon took us into the area of the shop floor, that the man on the shop floor increases our options. I feel that if Ken, together with the Trade Unions, together with the machine designer, was present at a meeting, the next meeting, I think that would again influence our thoughts. I feel that Ken's presence influenced our thoughts in the early part of the meeting, certainly influenced Mike, myself, and I think John, because John referred to it at lunchtime where we talked about the man on the shop floor in more detail. Just to sort of compromise, to what we're saying, would that be the next step?

HR   Well that's entirely for the committee to decide.

JF   I feel that's very desirable in principle. But I think that we've talked about seminars for the future and I would have thought we should choose our time very carefully, like when we've got something to say, we've got at least some initial thoughts reasonably ordered. But to jump the gun, and actually to take on board pressures to start doing design, which is what I think would happen, when we're still thinking about research, I think that could be premature. It's a matter of getting the timing right, but the principle I would agree with.

MC   But the division between design and research here, I think is a very subtle one. We nearly succeeded in designing back on the handle [words missed] this afternoon, and that option is rather subtly closed off. So I think, you know, it's a very subtle interaction between design and research, that's one area.
     My experience about industrial workers, in involving them in issues like CAITS and elsewhere is that if it's left purely at the level of discussion and debate and that sort of thing they're at a serious disadvantage, because I think our society does emphasise linguistic ability more than intelligence, [laughter] and I see great distinction between the two. Whereas if you actually get down to doing something, then you see what the amount of knowledge that industrial workers have and I can see that they would be probably; well some of them, I can think of one or two who would not, and therefore they would be articulating concern to the roof but we couldn't say that that is, you know, that this represents the views of a hundred people. It would be somebody who had thought about these ideas over the years and would be articulating them on their behalf, and therefore there is the problem of interpretation. But there are shop stewards on the shop floor and in particular people in prototype departments, where they're building prototype machines, who question every step the designers are making, I know that myself, and can contribute massively to the design of the equipment, so one could find people like that but even they, I think, might feel themselves at a bit of a disadvantage in this kind of discussion. But when it will come to actually making it, you know really sort of down there in your place and saying, well now, are we going to do it this way or that way. I'd have thought they could more than hold their own.

LK   It's so impossible to find a Trade Unionist. That would be super, but we've tried, haven't we?

HR   Well, not really, no. I think we haven't really said we want a Trade Unionist, let's go and find one. I think we've rather said that there is a problem keeping in touch with Trade Unionists and one way is to bring them in at a meeting that we hold, a seminar later on, and so on.

MC We have had the German Unions here and I'm feeding out more material when it became clear that we had something to feed out, and there's been the article now in Drafting and Design. So it's beginning now to be something more tangible, you know, because there is a danger that we'll be talking about something that doesn't really exist. Whereas it's reaching the stage now when it's all becoming more firm and concrete and I think we are reaching a stage where we'll have to make decisions about the degree of involvement.

HF We have back in Trondheim had several projects, research projects, involving Trade Union leaders [sc. shop stewards] doing actual research and we find some of the problems which you have mentioned and also with the job of reacting to concrete things rather than voicing things up in discussions. I'm sure there is a way of bringing this point of, let's say the optimum point when one could bring in Unionists or [words missed] . . .

My other comment is it seems to be a concern that if we bring in a worker or Trade Unionist then he is only going to represent himself. Now it seems as if everybody here is only representing a very small part of their own central community, and in the respect that this is an off-the-main-track project, you know, the engineers try not to think as engineers, nor the AI man who wants AI to produce, to enhance human skills rather than what I think seems to be the main track of AI research. So I don't really see this problem of an ordinary worker who presumably might only be voicing, or raising his own voice. It sounds as if, once we get to work [words missed] he has to stand for what everybody does and we have to constrain our solutions to what he says, which is not true.

MC I was wondering how in practice this would work out. Would somebody, a skilled worker, be working for a time with the team? And how could we arrange that? Would somebody be seconded from some company or would we have to find funds or could we get some unemployed skilled worker – unfortunately there are many of them about . . .

? That's true.

HR Yes, well that's something that, you know, would be the next problem, how would one do it? The first question is, is it a good thing to do? The point which I read into what Ken said when he came earlier, when he hammered at me much harder in fact than he's hammered at the Committee, [laughter] was that we're perhaps erecting our own professionalisms of various sorts as a barrier between the project and the shop floor, so to speak. Each of us is a professional: engineer, social scientist, computer scientist, and so on. And we're tending to say perhaps that only professionals can do design, and this is going to then be used for the shop floor. Now, of course, we are all nice people, we want to make a good system for the shop floor. But don't let them get near the design [laughter].

LK Look, I do think this has to be scotched. And particularly if that accusation comes from someone, ab initio, without having seen what we do . . .

HR   No, no . . .

LK   and it's purely from prejudice . . .

HR   It wasn't Ken's. It wasn't Ken's. It's what I read into what he said to me. And I don't think Ken was saying . . .

LK   As a social scientist my concern is reality: that people who are on this committee or in the team are there for the reason that they are alleged to be there and not for some phoney reason. If we include shop floor operators either in this or in the team, it has to be because of the contribution they are making to the work. Whereas what we are liable to do, if this discussion goes on much further, is to take them on in order to have a defence against that sort of accusation. And I would really very strongly resist taking people on in a PR sense in order to have that kind of defence.

KG   So would I, but I would . . .

?   So would I, yes.

LK   But they're liable to do that . . .

?   I don't think we're at all liable to do that . . .

KG   I don't think anyone would argue that position at all. I mean, what I would be concerned with though is, is the point that Håkon was making about machinists actually being in a situation where they are reacting to proposals that, that people are presenting to them, you know, options that they are to react to and I think it, you know, we have to consider the alternatives, of actually workers making proposals themselves within the project rather than continue reacting to proposals that other people make to them. Now if we can create a design team at some stage where that's taking place – I take the point that was made about workers feeling much happier in a situation where they don't have to depend upon their linguistic skills to make a point. You know, where they're actually working with things that they're doing every day. They're working with machines and developing machines or pulling machines to pieces . . .

LK   But if I could just finish a point, you know, the notion that – all right, I don't know if it's Howard's words or yours – but only designers can design. The skill of a designer is harnessing the contribution; that's what it's about, the skill is the methodological one, not the business, the way to do it, but how to harness the input.

HR   Can I, can I just interrupt . . .

LK   That is what some of the social science role is going to be.

HR   Can I just interrupt. I have to go because there are other urgent meet . . . matters [laughter]. I have to go to a meeting. I won't stop the discussion, I would like to ask Harold if he will act as a neutral referee. [laughter].

HP   The one who knows least about it. [laughter].

HR   So, I'll leave it with you.

LK   Are you coming back to us or not?

HR   No, it'll be two or three hours.

JF   Next meeting? After you come back from Australia?

HR   Yes, surely. Obviously. And my secretary will 'phone around. [HR leaves.]

HP   Well, I'm sorry about that Ken, I think we are onto an important issue and it's a pity we've sort of squeezed you right down to the end. I think Lisl you are still making a point.

LK   Well, I've made it.

HP   You're happy?

?   [Several speakers simultaneously, transcription not possible.]

JF   Ken, have you accepted this distinction that was made . . . [tape faint] . . . perhaps there's some consensus about, the distinction between this sort of fumbling stage and the working with the designing group. We're still in the fumbling stage at the moment. Perhaps you don't accept that. If you don't, then Håkon's . . .

LK   I don't.

JF   Well . . .

KG   It is difficult, isn't it, because . . .

JF   And I think that we might make a point that that distinction is a subtle one, I think whether it's settled or not, it's a matter of the weight of activities, and the activities at this end of the project, the [voice fades] very different, the whole project will be changing all the time. And, but Håkon said, can we bring that forward? Bring the time at which someone might be introduced. And it seems to me that the constraints [voice fades] the practical constraints of finding someone, at whatever stage, are going to be enormous. And finding somebody who is available, who has experience in the right sort of industry, the right sort of background, and is used to expressing himself in whatever terms – not going to be in awe of the people, a bunch of egg-heads like us, a lot of other things you're going to find, it'll be very difficult . . .

AC   John, we're all, we're quite a long way from that . . .

LK   John, you're ruining it, you're ruining it . . .

AC   We're quite a long way from that . . .

LK   The issue is a totally different one.

JF   I'm sorry . . .

AC   I mean when someone comes along as an observer with Ken to make a contribution within one of our meetings, and then we'll take it from there, as a Committee, as to just where we go to next, what sort of contribution can be made, is there an understanding at Coventry and so on and so on. This is what I was looking for as the next stage.

JF   Yes. But [indistinct] perhaps you could fix the task at the moment but I mean, the distinction . . .

AC   Just as Ken's come along today to make his contribution . . .

JF   Somebody else will come along . . .

AC   Right, somebody else from the Trade Union will come . . .

JF   . . .a machinist at some stage, a question of . . . I know you . . .

AC   John, not yet. We don't move into that stage.

HP   Allan, it seems to me that there are two distinct parts of it, one is should we try to involve operators in this Enabling Committee – do you call that an Enabling Committee? – that's one activity and then there is the research team. And it seems to me

[End of tape. Gap while tape is changed.]

JF   Well, I'm dealing with the fact that there are lots of skills, it would seem to me to be an advantage, there are many skills which are going to emerge, creative things, and that at the moment – but please make the point about what you see as the alternative.

KG   I don't wish to ruin anything . . .

LK   Oh just shut up a bit. [laughter]

JB   You've got to go in the same car? [laughter]

MC   He might [words missed]. [laughter]

LK   The model being put forward, is that here is a paternalistic bunch of people who claim to know best what operators need, and I am arguing against that model, whereas you are confirming that model.

JB   I think what John's saying . . .

LK   What I'm saying is two things – (a) I am not in that model, I, I resist that charge because I consider my skill to be in how to harness the contribution. And secondly, I think anybody who attacks me for this may be saying that *he* knows better what operators need. I have met this many times before. And every time I try to pursue this, John actually confirms your suspicions. [laughter]

JB   Does anybody else think – this is slightly nearer [voice fades] – we seem to have two types of design we've got buzzing around, we've got a little computer, there's designing for the operator: how are we going to use this equipment, which is what we tend to be talking about in the use of the machinists' knowledge. And there's actually designing how we are going to do it, there's how are we going to actually solve the problem that this machinist has helped us decide what we want to do. Now that, that second design phase is where the real, like, the real technical software expertise and the software scientists and, and wouldn't really involve the machinist because, if you like, his contribution has come before that.

We want to create a system that is a, means something to him and he's happy with himself. Now I think John's point that he's making is that we have got a number of techniques we are going to be using which are the actual detailed design of the system which is very difficult to imagine how the machinist can do anything other than say really I want to see it. Sort of asking me this sort of question and this sort of procedure, but I don't know how you do that. The machinist would say I have got no idea how to do that, [word missing] I'm afraid, but you know it really is true. And what we seem to be talking about is how we are going to take account of the machinist's idea of what this grandiose conglomeration of equipment is going to be treated, from his point of view, and I think my view now is that it's an, it is an earlier stage of design than I had appreciated. Because we're not at the stage where we are going to actually start writing coding and get the thing to work, we're at this fumbling stage, as John Fox said, and unless we take account of what the operator thinks in this fumbling stage, we *will* fumble it, because we won't have taken his view into account. So perhaps it's, my view is now that perhaps we should take account of it from some intermediate stage.

AC   What we talked about earlier is that we were going to do some research with two separate layers and we're going to look at the various options of [word missed] the two separate layers.

JB   Oh, you mean earlier this afternoon.

AC   Yes, and we were going to possibly bring along some operators from Long & Crawford [AC's Company]. So that deals with the operator side, working alongside social scientists themselves. Are we now looking at the Trade Union side, should a Trade Union representative be here? We can deal with the operator side through Long & Crawford. Someone coming along after you've done your research. In other words, let's not go back on what we've already said this afternoon.

JB Yes, that's right, I was fairly happy this, what was he called, the Research Officer out of the Union. I'm I don't know what, not interested in his qualifications, but one would expect a Research Officer out of the Trade Union, to be able, if you like, capable of understanding what we're trying to do and putting his, if you like, just his Union's viewpoint.

LK Actually I'm, I'm not sure if the, the Research Officer's the right one they're quite often in a very restricted position.

MC Yes, often they wouldn't be acquainted necessarily with the techniques on the shop floor. Unfortunately some of them are economists . . .

? [Several speakers simultaneously, transcription not possible.]

HP John, I think you and I now are talking about getting a contribution from skilled operators – as you're doing your research, and your design, and the final evaluation; and to me the operator's skill is important all the way through that and, I think in a way we're talking about how you tap that skill.

JB I think probably the point I could make is that we will only have the chance to do one design, I think, that's the trouble. And let's say that we've a typical design spec, comes down on several pages, in that spec we've got to have this viewpoint expressed. But when it gets passed over to whatever expertise we've got in the team, we are, if we're going to achieve anything, actually, sort of have an end result – and by that it might only be papers – we're not really going to be able to say, well as we're going along through that actual sort of handle-turning process of getting the results out, the work done, to be able to chop and change the design as we are going along. It's a bit like a PhD thesis, that you spend the first year sorting out what you're going to do, you spend the next year actually doing it, and you spend the final six months turning out the papers and thesis. And I do see it in those terms, the machinist's viewpoint has to be taken account of: there comes a stage where we've actually got to do the work. And I'd be worried that if we start fiddling about with what we're trying to do, then we won't actually achieve anything. And by all means let's evaluate it afterwards, and I do agree with that.

? [Two voices at same time, transcription not possible.]

HP Allan, you need to go?

AC I've got to go, yes. I have a car coming down at four o'clock [voice fades].

HP Well, we seem to have reached a point of agreement that we need to tap the skill of the machinist. And we don't . . .

AC Well, I would define that as we agreed earlier when we've done some more research. What we were talking about was representation around this, in this Committee. Do we move it to asking Trade Union representatives to come along? Were you suggesting that, Ken?

KG   Well, on one level. Because there are a whole series of, of questions that can be raised about that. I mean on one level, on the Steering Committee, a Trade Union representative is, I think this is very useful to have. A Trade Union rep with those sorts of experiences of, of the systems actually in discussions. I think that would be very useful. Not someone like a Research Officer who's got economic skills, but, someone with practical machining experience.

JB   I'd rather opt to drop these Research Officers . . .

KG   So to answer that question on a practical level, yes, I think it's useful to have someone on the Steering Committee. But the other question really is all about the design team, that aspect of it. And my feeling is that it's not adequate simply to, to call upon your workers at Long & Crawford to, to actually be involved in a discrete piece of research and then, you know, the thing goes on, perhaps moving to another firm or moving to another part of the country to look at another aspect, you know, and then coming back to Manchester again. My question is, isn't it possible to involve machinists on a continuing basis through the project in, in the design work?

HP   Well certainly John, the early papers that I saw, there was a lot of debate about how do you find typical operators, I think, and how do you involve them in the, in the development work, and I'd assumed that you would be working that out as you went along.

JB   Yes, I think, I, yes, I think certainly I'd always assumed that we would be, if only getting the reactions.

HP   Yes, yes but I think Ken is, Ken is suggesting that you try to do it in some sort of systematic way rather than just let it happen. Is that . . .?

KG   Instead of on discrete pieces of work, calling on . . .

HP   Ongoing?

KG   Ongoing so that someone is there as, as part of the team.

JF   I'm sorry, I keep on being counter today, but I'm not entirely clear, this person due over the next few months [voice fades] there will come a point where that person can be valuable [voice fades]. But we just, all I can do is say I'm drawing on my own experience, and I'm sorry to hear that [voice fades]. But my work, with software . . .

LK   Well, it's . . .

KG   I think, I think that, that there is that charge of paternalism.

LK   Yes, I mean, that's the charge that's being made to us.

JF   [very faint] . . .don't really care, actually [voice fades]. I've worked with many people, who – my role could not be perceived.

LK   [at same time] I also don't think it applies, that's what I'm saying.

JF   as paternalistic, could not be regarded as paternalistic. It could be regarded as service . . .[voice fades]. In that situation when there are stages in the sort of work that I mean, where there is design, and I have to consult partly by research. But at an earlier stage, I wouldn't dream of talking to people – I have actually been burned, they're obstructive, they, they present views which are beside the point, because they have no – when you are dealing in a new area, it's the technical points. I take the charge of professionalism, maybe exclusiveness but not paternalism, when you're trying to sort out a set of technical problems wet and new, not recognised, not what other people suggested and what have you, that there is a problem to be solved, and you're trying to say look, how could we solve this, conceivably solve this technically? Then you, you really want a quiet moment, maybe for working in committee, then you try to bring the requisite skills together to solve the problem. But to go back to the customer too early, or the client or whatever non-paternalistic terms you want to use, can be very destructive, and I am not in any way opposed or arguing against settling the date at which one goes back to the client. I am simply saying that at the moment I don't think we're in a state to offer anything useful and to bring in people who have non-technical views about the problems of production. Now I realise that may close things off.

HP   So do you want to make a last point?

AC   Yes, can I make a point then. I think John should think very seriously about this and report at the next meeting, personally.

HP   John Boon?

AC   Yes, this John. John's, John's got the problem.

JB   Can I just ask Ken something. A couple of questions.

HP   Sorry, Lisl's been trying to – I was trying to let – OK, good to see you. [AC leaves.]

LK   I have heard a great deal of the sort of lines you're taking and I don't know whether I have conveyed my objection to it, which is that it's too easy. If we did everything you say, and if we did it much more than you say, like we all withdrew and the whole thing was totally participative – there was nothing but machinists on this Committee and in the design team – within six months the charge of paternalism would be made to them, because whenever participative design has been set up, what has happened is that you have created – if you don't understand the dynamics, and work with them and work it out – you've created a new élite.

KG   Well possibly new élite would be a charge but not paternalism. If those people . . .

LK They will be designing for others. The moment that the guy who works the eventual system down the road, once people have bought it, isn't the same as the guy who designed it, the problem is this.

KG But if those people are coming from the same background. If there's . . .

LK It doesn't matter. They come from the same background but they're still not the same people.

KG I would, I would really question that, because if those people are coming from an industrial background, where they are in a situation where this machinery is going to be used by the employer – and the employer's particular interests and the Trade Unionists' particular interests in how that machinery is used – if those Trade Unionists are involved in the design, design of the system they are considering those interests when those designs are taking place. Whereas I think social scientists can at times divorce themselves from those interests because it is not an objective reality for them.

LK They're still not the same people who are the users. The gulf still exists.

KG Not the same gulf.

LK Yes, the same gulf.

KG Well, I wouldn't accept that at all because the social scientists' own material reality, their own material interests, are going to be very different, they're pursuing different careers. They are relating to different people . . .

LK So are your people, the moment they be – they join this team. So are your people the moment they join this team.

KG But they'll only be on the team for two years and then they have to go back and report . . .

LK We're only on the team for two years.

HF [short interjection, inaudible.]

KG Yes, but then your career after this project would be different.

LK Theirs will be changed.

? [Loud voices simultaneously, transcription not possible.]

LK So will theirs be. So will theirs be. How many Trade Unionists who go to Ruskin go back into the Trade Union movement?

KG Very few.

LK Right.

KG But we are talking about machinists who are involved in a project on developing a machine system for two years, that's, well I mean, the possibilities of enhancing their career in the Trade Union bureaucracy are there I suppose, but *extremely* limited.

LK They will be asking for research jobs in UMIST, that's what it is.

HP You have also got the problem, the same problem, of acceptability of the system at the end, haven't you, to, to the ordinary machinist. Håkon's been trying to get in edgeways.

HF I don't think you can. I think you're right in this, Lisl, you can't, you can't bring in everybody who's going to use this system at, at the, in the end, and so you'll always have the – call it paternalistic or in similar words – charge. Ken is right I think, though, in the sense that people with a different background will bring in different, different interests into the project. It is possible if they're allowed to. And if they also have the skills of promoting their own interests. And one, one thing that people could bring in, which not necessarily would be brought in by social scientists, would be the combination of, of knowing the machinist's trade, with the knowing of the industrial relations game. That is, how is the piece of machinery going to operate in the total industrial environment. Now that's an input that I can hardly see, or I can see very few social scientists being able to bring. And your point comes in, OK this is something we can, we can get from these people by questioning.

LK No, on the contrary, we can't do this at all.

HF Well . . .

LK We have no control how the thing is eventually used.

HF No, no well in the end . . .

LK It is not part of the design.

HF You can't get it all out. But your point is that the reason for the specific design skill that professionals have, that consists in, sort of asking people, getting on the job and finding things there, with experience. And you say this is, you say rightly that this is a professional skill that one has to learn. But then it's the professionals have set the agenda of what are the, what are the important things to put into it.

LK But the only design skill I have – and, I'm not in the project team, you know, I'm not the researcher – the only design skill I have is working with simulations and models, working with long lists of design criteria, and trying to get people to evaluate, to give their importance weighting to, and then getting that importance weighting tested out on alternative designs. That, that is about policy, it's not about asking.

HF   No, I'm, I'm sorry that's a language problem.

LK   Okay. I have absolutely no skill in saying a job should be set up this way or that way. I have personal preferences, which I can make clear.

HF   Well, I wasn't implying that you're a, that you are in a position that you think of yourself as being in a position of designing a job. But I'm perfectly aware of the issue, from the designing, participants' design of system.

LK   Now how the thing would eventually be used, the total industrial relations environment, in which this thing will eventually be used, is out of our control anyway.

?   [Several loud voices simultaneously, transcription not possible.]

LK   It is not part of the reason of this project.

KG   I raised that as a question actually, because I would've thought that an important link is there, between Trade Unionists and the project. As I was saying before, I thought that an important point about this project is that it was offering to Trade Unionists a positive proposal that they could make in collective bargaining about technical change. So that, I mean, far from saying we have no control over the situation we, we clearly, although perhaps have no ultimate control, clearly want to influence the situation of collective bargaining by actually providing Trade Unionists with something which we think could be useful for them to actually consider when they are considering their response to management's decisions over technical change, so that we are actually involved in, in that process. It's not something where you are actually designing quite separately and then letting the market judge or letting society choose. We are actually making a political input.

JB   Yes, I don't think we are actually offering negotiating chips to the Unions. I think what we are doing is offering an alternative to the management system, if you like. I don't think we've ever said that it's a thing that's in essence being designed for a Union to use, it's an alternative approach which does seem to be more attractive from a Union point of view. Now I've been, I think we would also think the Unions say, Ah, but this is a much better way to do it, and if you like that's the reaction we want to get from the Unions or the machinists, because I think the two are the same, I wouldn't like to distinguish between the machinists and the Unions.

HP   Ken, Ken, could I ask for a point of clarification. It, it seems to me that, that there are two issues being talked about now. One is to get a better system at the end because the operators have been very well involved. You seem to be also raising the question of strengthening the position of the Trade Unions in collective bargaining. Is that, is that fair?

KG   [Assents.]

HP   Now I don't really see that you can use a research project of this kind, in that sort of political way, for the Trade Unions.

MC   Well, I wouldn't have seen it in, in such overtly political terms but I've always been concerned that confronted with technological change, Trade Unionists are driven into a very negative response position. They've either got to say they're not having it and they're accused of being Luddite or they have to, have to accept a wage increase for what ultimately becomes either the displacement or the de-skilling of their members. That may be now vulgarising it but I think in practice that's what it seems to amount to, and we were impressed with the experience of the colleagues in Norway and elsewhere who didn't just say that is the only area of legitimate industrial relations negotiation, there is the whole question of what type of system, how the system is designed, do we have the right to participate, so I saw it in that sense of strengthening the position of the Trade Unions, in other words opening up the area that is legitimate for industrial relations.

HP   Playing a positive role.

MC   Yes, yes.

HP   But my personal experience is that the Unions do play a positive role, particularly in a project of this kind. And my experience is that in fact Trade Unionists have a longer-term view, a much longer-term view, than the immediate managers who have a very short-term view. And that the view of the Trade Unions is, is this going to protect my job and my grandchildren's jobs and they are, they are in favour of, of developing better technology in my experience.

JF   Can I say that as far as the two issues in separation are concerned and I'm glad you brought that out because I don't know whether you can really quite – you added the phrase 'frankly political', something like that.

HP   [Agrees.]

JF   Certainly, if there was any alliance between this project, between an end that was frankly political whatever it was, political advantage, I would regard the project as compromised and I would actually resign. However, that leaves aside the question of whether people have specific contributions to make, and it seems to me that everybody round this table is agreed that there is, there will be some known, some unknown, skills and contributions that somebody who doesn't have our experience, has other experiences, will be able to make, and that's a great weakness. The debate we're having is about one of implementation it seems to me and professional protectionism, what have you, I think charges like that seem out of place, it's a matter of practicality and it might be with the decision, after a debate fairly had, that we should do it tomorrow, you know, but I'd want to have that debate and at the moment I am not convinced that it's appropriate to enlist, or to elicit contributions of a technical, semi-technical nature from the workplace, I mean technically involved [word missed] right now. I don't think we know what sort of contributions are appropriate. That's the position.

HP   I think that puts it very fairly. Do you want to say anything Ken?

KG   No, I don't. Pass.

HP   Håkon will speak to us. Sorry, do you mind Lisl. Håkon.

HF   Maybe I should emphasise that my interest in the decisions has been from the professional point of view that concerns the relationship between the designer and the designed for. We have been doing some, some research in Norway and we have been trying out participative design. And we are trying to involve the users at an earlier and earlier stage. We also have quite a number of projects where Trade Unionists are on the steering committee, and I, I think there, I think there is a possibility to find a role for, for a machinist with a Trade Union view, or vice versa, on the Steering Committee. Maybe also, possibly, on, on the actual project team.

HP   Håkon, could I ask a question? I see John's point very strongly, that when you are in the, the idea forming stage, to talk to the wrong person is very damaging. It, it doesn't matter whether it's a Trade Unionist or, or a, sociologist, you've got to be able to have a positive discussion with people. Is, is this a problem in Norway – finding the Trade Unionist with the right kind of positive, forward looking attitude?

HF   I don't think the problem is of finding a particular person with the correct attitude, rather the more structural things like, there is barriers of language, there is barriers of understanding. And we, maybe this Committee is already so, so conglomerate, that adding still another viewpoint in here might [voice fades]. Maybe that's what the debate is all about.

HP   [Has assented all through previous speech.] Well, my experience . . .

LK   Could I say . . .

HF   As I say, although we haven't been able to find a very, a very good working model for involvement [voice fades] but we have work in progress and I think it is possible to involve people at an earlier and earlier stage.

HP   Lisl has been trying to . . .

LK   I may not have conveyed what I'm about. I am *so* committed to participation that I would fight phony and pretence participation. If we were designing in a company for something that's being implemented in a company, I would have fought for the involvement of – however unskilled and however inarticulate people – way, way, way back. You wouldn't have got anywhere near as far as this without. But that isn't what we're doing. The design here is of particular equipment which companies will then buy, and your token machinist isn't going to meet the need for participation. That sort of participation is going to have to happen and we haven't yet talked about, if you like the instructions that go with the equipment to the purchasers, that's where that kind of, well, your real person, however inarticulate etcetera, needs to have power at the implementation stage.

KG   At the implementation stage?

LK   And what you need at this stage . . .

KG   What is the implementation?

LK   When somebody's actually installing the equipment.

KG   Well, I think that, that all the decisions have been made then.

?   [Several voices together, loudly, transcription not possible.]

LK   Hold it. At this stage what you need is to be very careful that that person can then exercise power, so that the right options are left open at this stage, that's what will happen, and your single token machinist on the team isn't going to do that for them.

JF   Well, I agree with that. Can I just, just add, and I don't want to stop you answering, the, the point about distinction between [word missed] the boundary is blurred and hard to identify [word missed]. That has been made many times and accepted. [Words missed] the thing is to fulfil the perspective that this has given us, but it does seem to me, I mean, I deal with people who are really not into – in my work I deal with doctors. If I involve them too early they – I mean you basically have to [word missed] cooperation.
  We come to this point of thinking about the project, after a lot of preparation, a lot of thought, a lot of accidents, and so forth, but we have a certain sort of common experience, and so on.
  Go out to other people apart from the technical people, go out to other people, articulate or not, they, it takes a long time to bring them out and you can't guide them particularly in the early stages because you're not articulate yourself, that's the point. You can't guide, you can't often, you end up with these, these destructive, posturing sort of encounters in which nobody's got [word missed] the point they want, and very often the project can fizzle out as a, as a result. It isn't actually stable at that point, you know. And that's, that's what I am trying to stick, to defend on. It's not, it isn't a professional point at all, it's a defence of, of – well, I've made my point.

KG   Well, I would, I would accept that sort of, I know, I think that obviously it would, it would be for the Steering Committee to judge when you felt that you'd got coherent ideas to begin to argue about with people who'd be using the kind of systems that the Committee would hope to see developed. Certainly what I wouldn't like to see is a situation where UMIST were developing a product for a market which then workers actually . . .

JF   I don't think you know Howard Rosenbrock.

KG   OK.

JF   I don't know him very well, but I know him well enough to feel that is, just is *not* what he's at.

KG   So it brings the question about when?

HP   You've missed an awful lot of it. Do you want to make any points?

?   No.

HP   Could I suggest we finish at quarter to five?

?   Yes.

?   Before the rush hour.

HP   Yes, I've got to drive a hundred and fifty miles, too. What about you, Ken?

KG   Yes, I've, I've got to go as well.

HP   Yes, is quarter to five all right?

?   [Several speakers assent.]

HP   OK. Sorry. Now then, Mike, do you want . . .?

MC   Well, I think it is the case that if you take people out of their environment and they're elevated to researchers in the way you described, they do become separated from that group and they're regarded by the group then as experts. And they're treated differently and so on. There has been the experience in Birmingham, at Lucas Aerospace, when new technology was coming in, a group of people who were going to be affected by that did elect a number of people to negotiate on their behalf. So I mean it was more in an industrial relations context, but within that they were discussing what the options were and arguing for them and then reporting back to that group. They were delegated in that sense, and they, at the end of the year they'd be replaced by others and so on, so to that extent there was that kind of safeguard. But it's incredibly difficult to then set up something like that in a project of, of this kind, and I think that at the early stage, when we are not even clear ourselves, that it may just add to the confusion. Although, as I said – there's no point in going over it again – that I think we do make design decisions here in a very subtle sort of way and that quite early on, if we are going to have this involvement at all, we should have it. But hopefully when we do, we'll ensure that there's a large component of practical issues, rather than just at the linguistic level, because workers will always be at a disadvantage at that level, whereas we will be at a disadvantage when it comes to actually doing things, that's my experience.

HP   Exactly.

MC   It kind of balances it out, the subtle stage at which we do that.

LK   For example, I wouldn't at all mind anybody listening to the tape and all our discussions being recorded, and saying Oy, there you're actually closing an option, or exercising somehow, inserting some prejudice without knowing it. I, I think we'll be totally open to that sort of thing.

HP   Well, it does seem to me Ken, that you, you are a sort of unique representative from, from Coventry Workshop, and to have the relations with practising people which could be very helpful. And . . .

JB   Is there any way that perhaps I could come down and see some of these machinists [words missed] prototype development work, not necessarily me, but so that we can get an appreciation of how they can contribute. I think, if you like, I think we've probably all got in our mind an idea of what a machinist can do and where he perhaps can't contribute and I, I happen, I, I think I know a chap . . . who used to be a designer in Herberts [word missed], the company I think you're talking about. And I know another chap who helped design controls for the MDI system. Now in other words the impression I got of those people is that they would perhaps create something and get the reaction which is very much, if you like, the way that it was done in the past, and it's done at the moment. Now if we can approach those same people and say that you know, perhaps I can sort of give them an explanation of what we're trying to do, but see if there is a way that we can, you know, outside of the meeting, for the time being let's get some idea of what their capabilities are, from my point of view or the Steering Committee's point of view. But let's get an idea of what they can contribute and then we're in a better position to judge when we can sort of say yes, let's get their contribution now, because obviously we can't get the actual information that we want out of them at the first meeting, I don't think, there's no way they can instantly come up and say we can't do it that way because – they're going to take a while to be involved.

KG   Well, I couldn't give you a yes or no on that because obviously . . .

JB   Well, that seems one way of getting, you know, making some progress on this issue.

HP   Yes. Have you any other, alternative suggestion?

KG   Well, you see, I'm a, I'm from Coventry Workshop, not from the Stewards' Committee, so obviously, I mean, I raise that from the Workshop's point of view, because working with Trade Unionists we think that that would be a useful input. So obviously what I could do is then from the Workshop, approach the Stewards' Committees and, and say you know that this project is, is interested in that sort of communication and I should think it'd be positive, but obviously I'd try and . . .

JB   It does seem one way of making a bit of . . .

HP   Well that seems, it seems to me an excellent outcome, but if, if you can ask your colleagues Ken, and then we ask John if he can report back to the next meeting.

JB Well, somebody's sort of stuck my name down saying we must answer this problem [laughter]. So I thought, well, if I'm going to answer this problem, I need . . .

HP Yes.

JB . . .to get some sort of feel.

HP Is that, is that reasonable, are people happy? You happy Lisl?

KG So you would want to discuss with some of these workers . . .

JB Well, I think obviously I could spend quite a lot of time, I could spend all day if I had to talking about what we're trying to do but I would be really much more interested in is sort of saying this is what we're doing but, you know, would you think there's any other way we can tackle it, you know, trying to get this sort of positive reaction. The action rather than the reaction is what I am looking for, if you like, which is, I think, what we're all sort of concerned about, that it's – and let's hope it works, it works.

HP And you, you happy to go along?

JB Ah, I think so, yes I . . .

HP And are other people happy for John . . .?

JB I, you know, I don't know if anybody might want to come with me, perhaps. It does seem something that we can do.

? [Several voices together, transcription not possible.]

? Yes, and it would sort of prepare a set of notes or something . . .

? [Several voices together, transcription not possible.]

MC And clearly in Coventry there's a wide range of views about these things just as there is amongst us, because Phil Higgs takes one viewpoint and the people in Alfred Herberts take another. Others still say the key issue is that the operators do the programming, whether it deskills them in the operation or not, so I mean, you know, there is . . .

HP OK. Well, thank you very much Ken, you have obviously stimulated a useful discussion. I suppose we ought to talk about the next meeting. Mid-September, is, how, how does that fit John, for the research team?

[End of this part of the discussion.]